T0302211

Understanding Cybersecurity Technologies

The Human Element in Smart and Intelligent Systems
Series editor: Abbas Moallem

This series aims to cover the role of the human element in all aspects of smart and intelligent systems. It will include a broad range of reference works, textbooks, and handbooks. The series will look for single-authored works and edited collections that include, but not limited to: automated driving, smart networks and devices, cybersecurity, data visualization and analysis, social networking, smart cities, smart manufacturing, trust and privacy, artificial intelligence, cognitive intelligence, pattern recognition, computational intelligence, and robotics. Both introductory and advanced material for students and professionals will be included.

Smart and Intelligent Systems: The Human Elements in Artificial Intelligence, Robotics, and Cybersecurity
Abbas Moallem

Understanding Cybersecurity Technologies: A Guide to Select Right Cybersecurity Tools
Abbas Moallem

For more information on this series, please visit: https://www.routledge.com/The-Human-Element-in-Smart-and-Intelligent-Systems/book-series/CRCHESIS

Understanding Cybersecurity Technologies

A Guide to Selecting the Right Cybersecurity Tools

Abbas Moallem

CRC Press
Taylor & Francis Group
Boca Raton London New York

CRC Press is an imprint of the
Taylor & Francis Group, an **informa** business

First edition published 2022
6000 Broken Sound Parkway NW, Suite 300, Boca Raton, FL 33487-2742
and by CRC Press

4 Park Square, Milton Park, Abingdon, Oxon, OX14 4RN

© 2022 Taylor & Francis Group, LLC

CRC Press is an imprint of Taylor & Francis Group, LLC

Library of Congress Cataloging-in-Publication Data

Names: Moallem, Abbas, author.
Title: Understanding cybersecurity technologies : a guide to selecting the
right cybersecurity tools / Abbas Moallem.
Description: First edition. | Boca Raton, FL : CRC Press, 2022. | Series:
The human element in smart and intelligent systems | Includes
bibliographical references and index.
Identifiers: LCCN 2021031432 (print) | LCCN 2021031433 (ebook) | ISBN
9780367457457 (hbk) | ISBN 9781032157849 (pbk) | ISBN 9781003038429 (ebk)
Subjects: LCSH: Computer networks—Security measures. | Computer security.
Classification: LCC TK5105.59 .M639 2022 (print) | LCC TK5105.59 (ebook)
| DDC 005.8—dc23
LC record available at https://lccn.loc.gov/2021031432
LC ebook record available at https://lccn.loc.gov/2021031433

ISBN: 978-0-367-45745-7 (hbk)
ISBN: 978-1-032-15784-9 (pbk)
ISBN: 978-1-003-03842-9 (ebk)

DOI: 10.1201/9781003038429

Typeset in Times
by KnowledgeWorks Global Ltd.

Contents

Acknowledgments

I would like to acknowledge the following people who have reviewed drafts of this book and provided valuable comments and suggestions. Without their support, this book would not have become a reality.

I'm immensely grateful to cybersecurity experts Wojciech Cellary, Professor Poznan University and Mohd Anwar, Professor of Computer Science at North Carolina A&T State University for their insightful, reviews, comments, and suggestions.

My deep appreciation to Shabnam Moallem who used her artistic talent to illustrate the analogies of each chapter.

I am grateful to Arash Baytmaakou who edited the text and improved it with his suggestions.

Thanks to Aparna Avinash Gokhale who refined the tree views of the technology figure of each chapter.

Introduction

Cyber-attacks on enterprises, government institutions, and individuals are exponentially growing. At the same time, the number of companies, both small and large, offering all types of solutions has been growing too. Since companies rely on technological solutions to protect themselves against cyber-attacks, understanding and selecting the right solutions among all those offered presents a significant challenge for professionals, company executives, and newcomers to the cybersecurity field.

At the 2019 RSA Conference, an information security exhibition that took place in San Francisco, California, over 700 companies exhibited their products in the field of cybersecurity. Does this mean that they are offering over 700 unique technologies? Obviously not. The number of real cybersecurity-related technologies is much smaller than the number of companies offering them. Many of these companies use the same technologies and bundle them differently for sale, sometimes making minor differences. For example, there are many companies offering antivirus solutions. Do they use the same technologies or different ones? If different, then what are those differences? If someone wants to select an antivirus solution, the most important criteria would be to understand what technological solutions are used in the product. During an investigation of cybersecurity technological solutions, I examined cybersecurity technology-related books, magazines, blogs, articles, and company websites for product descriptions. After spending significant time navigating, reviewing, and reading these resources, I have learned:

- Most of the books written by different experts are incredibly technical and not very understandable to all levels of cybersecurity professionals.
- Existing books are often out of date due to rapid changes in the cybersecurity field.
- Books are also often too general and do not include easy-to-understand descriptions of technologies.
- Technical magazines, articles, or blogs are mostly written by influencers, which means the information provided is not always objective or reliable. These resources often rank companies, solutions, and focus on providing general descriptions of the solutions.
- Companies only offer solution explanations with generic marketing materials without an in-depth, easy-to-understand description of the technologies. Sometimes the descriptions are generic and simply list the features of the product – not the technology used to provide those features.

After reading different companies' materials, at the conference, I decided to ask for a verbal explanation from each of the vendors at the cybersecurity product exhibitions. While listening to their marketing presentations at the exhibition halls, I realized that most people attending the presentations were struggling to obtain information. It was a very disappointing experience.

The presentations were extremely generic, even more, general than reading the material. Sometimes the content was also presented by people who themselves are not tech experts. The generic PowerPoint presentations were often accompanied by a few quiz questions that allowed you to receive a token prize.

After that experience, I researched different companies' management meetings about cybersecurity solutions and came to the conclusion that in many cases, only the security managers of their respective companies were highly informed and showed a strong understanding of the topic. The rest of the time, it was apparent that attendants were not well informed and did not have in-depth knowledge of the actual technologies.

Following this experience, I decided to challenge myself and create an understandable guide of cybersecurity technologies that focuses on the technologies that exist and are used in different solutions, without focusing on the companies that offer such technologies.

The first step is to provide reliable and easy-to-understand classifications and descriptions of each group of technologies.

This book is intended to help all professionals new to cybersecurity, students, and experts to use the content to educate their audiences on the foundations of the solutions already offered.

After a brief classification of the main technologies (Chapter 1), each chapter will focus solely on one type or group of technologies. The intention is to create concise and digestible chapters, avoid technical jargon as much as possible, and focus on open-source technologies. Each chapter's content is consistent: after an introduction and historical background, I describe how the technologies work, explain the main technologies within each group, highlight the main advantages and disadvantages of those technologies, list the products that use those technologies (when relevant), and end with a short conclusion. To help with understanding the technologies, I use analogies and a visual illustration (a treemap with different branches) to assist non-technical readers and to provide a mental image. Additionally, in order to provide clarity with terminologies, all technical terms and definitions used in each chapter are included in the glossary section, along with the names of people mentioned in the book.

Abbas Moallem

Author

Abbas Moallem, PhD, is a consultant and adjunct professor at San Jose State University, California, where he teaches human-computer interaction, cybersecurity, information visualization and human factors. He is the program chair of HCI-CPT, the International Conference on HCI for Cybersecurity, Privacy, and Trust.

Dr. Moallem is the editor of *Human-Computer Interaction and Cybersecurity Handbook* and the author of *Cybersecurity Awareness among College Students and Faculty*. He is also the editor of a book series from CRC Press titled The Human Element in Smart and Intelligent Systems.

Figures

1 Cybersecurity Technologies Classification

1.1 INTRODUCTION

Cybersecurity is a growing field of science composed of very diverse types of technologies used to protect systems from different kinds of attacks, some known and some unknown. Due to the complexity of the evolving aspects of these areas, there is no standard classification of protective technologies. The classification of cyber technologies helps to better understand how these technologies work. In this chapter, after a look at some of the different available classification, I will provide the way I have classified the protective cyber technologies reviewed in this book.

1.2 DIFFERENT CATEGORIES OF CYBER TECHNOLOGIES

There are different classifications of technologies in cybersecurity. From a research point of view, the cybersecurity topics are categorized as follows [1]:

1. Applied cybersecurity
2. Cybersecurity data science
3. Cybersecurity education and training
4. Cybersecurity incidents
5. Cybersecurity management and policy
6. Cybersecurity technology
7. Human and social cybersecurity
8. Theories in cybersecurity

From a practical point of view, one type of classification is commonly based on the kind of threats and their intended effects. In this category, threats are classified into two groups [2]:

1. Attack techniques
2. Threat impacts

Another type of classification is based on the type of cyber protection. The protective technologies can then be classified into four groups:

1. Architecture system
2. Detection type

DOI: 10.1201/9781003038429-1

3. Ecosystem
4. Data type

Cybersecurity technologies can also be classified based on components that each technology is supposed to protect. This classification includes the following main groups:

1. Network
2. Application
3. Endpoint
4. Data
5. Identity
6. Database
7. Infrastructure
8. Mobile
9. Cloud
10. Backup

Another type of classification is based on security solution types. The solution-based classification is a more empirical way to classify cyber technologies. In this classification, the solutions are classified in the following groups:

1. Identity and Access Management
2. Risk and Compliance Management
3. Encryption
4. Data Loss Prevention (DLP)
5. Unified Threat Management (UTM)
6. Firewall
7. Antivirus/Antimalware Solutions
8. Intrusion Detection System (IDS)/Intrusion Prevention System (IPS)
9. Disaster Recovery
10. Distributed Denial of Service (DDoS) Mitigation
11. Web filtering

1.3 TECHNOLOGY CLASSIFICATION

For this book, I have used classifications based on existing technological solutions. They are grouped into 16 main categories. In each chapter, I offer the tree view of the technologies under each leading group. The classification in this book includes the following groups (Figure 1.1):

1. Encryption
2. Authentication
3. Biometrics
4. Firewall
5. Endpoint Protection

6. Phishing Detection
7. Virus Detection (more commonly put together in a single group with the Endpoint Protection category)
8. Malware Protection

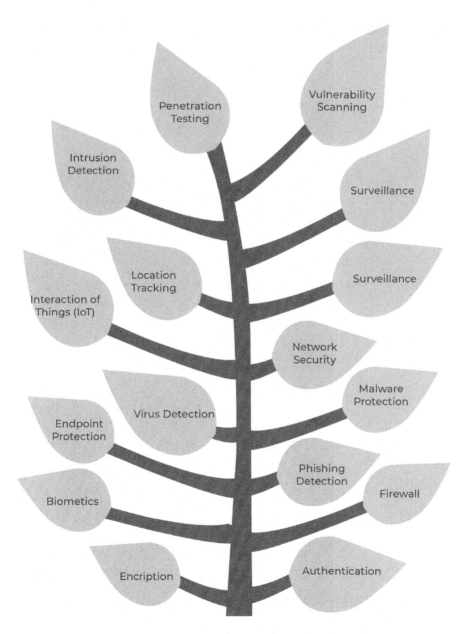

FIGURE 1.1 Cybersecurity technology classification based on existing technological solutions.

2 Encryption

2.1 INTRODUCTION

People use online technologies in daily activities. Web and mobile application companies get millions of megabytes of data every day from customer activities. The data collected from customers might include personal information like date of birth, credit card details, login credentials, etc. People communicate and exchange information, sign documents, communicate about confidential documents for business or private purposes, and they do this all through text, messaging applications, and email.

Organizations and individuals need to make sure that all data is securely transmitted and stored, and that the data exchanged cannot be viewed or accessed by other people or parties. If the data is not secured during the exchange, the cost of data breaches and customer trust can be monumental. In order to secure data in transmission, storage, or processing, the data must be encrypted.

Encryption is technology that takes plain text – like a text message or email – and codes it into an unreadable format. The science of encrypting and decrypting information is called cryptography. Unencrypted data is called plaintext and encrypted data, ciphertext. Thus, data needs to be encrypted before it is sent over the Internet. If encrypted data is intercepted in the transition, it will not make sense to anyone and the confidential details will remain safe because of the encryption.

Encryption is also essential for protection when data is stored. If passwords, usernames, or authentication credentials were saved in a database unencrypted, then anyone with access to that stored database would have that confidential information at their fingertips. For example, in 2012, Facebook confirmed that it had stored "hundreds of millions" of account passwords in plaintext for years [1].

After a brief historical background, this chapter will review how encryption technology works, the types of encryption that exist, the advantages and disadvantages of using encryption technologies, and the different products which use encryption.

2.2 BRIEF HISTORICAL BACKGROUND

The history of cybersecurity is primarily rooted in cryptography. Encryption was used for the first time in 600 BC by the Ancient Greek people known as Spartans. They used a tool known as a scytale to send their messages in an encrypted format during wartimes. A scytale consists of two wooden rods of

DOI: 10.1201/9781003038429-2

the same dimension and a leather belt with alphabets. The message sender used to send the leather belt to the receiver. The letters on the leather belt would not make any sense until they were wrapped around wooden rods with the same dimensions as those of the sender.

Later usage of encryption dates to approximately 100 BC. In this method (known as "Caesar Cipher"), Julius Caesar would send secret messages to his army by shifting each character in the message by three. For example, an "A" would be replaced with a "D," "B" with an "E," and so on. The characters would wrap around at the end, so "X" would be replaced by "A."

In 1467, Leon Battista invented the alphabet substitution technique. Later, Giovan Battista Bellaso, in 1553, made use of an encryption key for the first time in human history. The receiver would need to know the encryption key – agreed upon by both the sender and receiver – in order to decrypt the message. The encryption key became the basis of the encryption algorithms which we use now. In 1854, Charles Wheatstone invented the Playfair system. The Playfair system used two letters to encrypt words, which made them more difficult to decrypt.

In 1797, Thomas Jefferson invented the Jefferson Wheel. This wheel was used to send encrypted messages. The wheel had 26 alphabets on an iron rod in random order. Turning the letter on the rod would encrypt and decrypt the words. It was a reliable encryption technique for its time. The American military even used this technique from the 1920s to the 1950s and found success with it in both the World Wars.

Later, Edward Hugh Hebern invented the rotor machine for encryption. The rotor machine used a single rotor, in which the secret key was embedded in a rotating disc. The key encoded a substitution table and each keypress from the keyboard resulted in the output of the ciphertext. Each keypress also rotated the disc by one notch, and a different table would then be used for the next plain text character. This encryption was again broken (unlocking the code) by using letter frequencies.

After the First World War, in 1918 the German Arthur Scherbius invented the Enigma, an electromechanical cipher machine, replacing a pencil and paper system. The Enigma machine was used extensively during World War II by Nazi Germany military. Enigma was broken by code-breakers from Poland during the Second World War, who continued their work for the United Kingdom secret service within a secret program at Bletchley Park in the United Kingdom headed by Alan Turing [2].

In the early 1970s, IBM designed a cipher called Lucifer due to the high demand for more encryption technologies. In 1973, the National Bureau of Standards (now called NIST) in the United States put out a request for proposals for a block cipher, which would become a national standard.

In 1973, the United States started using the Data Encryption Standard (DES) to encrypt messages. DES was unbreakable for a few decades but was cracked in the late 1990s. After DES was cracked, the United States introduced the Advanced Encryption Standard (AES) [3]. AES is open source and is still used by the United States, proving to be a highly successful encryption technique.

2.3 HOW ENCRYPTION TECHNOLOGIES WORK

ANALOGY

To better understand encryption, think of it as similar to creating a secret language. You want to make sure only the people who speak the secret language understand. For example, imagine you want to send a secret message to one person or a group of people and you want to make sure that nobody else can read it. You might use simple encryption like what Julius Caesar did. For example, you could shift the letters so that an "A" would become an "E," a "B" would become an "F," etc., and then you would let the recipient(s) know the key through a separate communication channel. In this case, the encryption would be straightforward and modern-day hackers could quickly try many combinations to decode your message using a simple computer program.

Now imagine you are using a very complicated key using mathematical equations that code your message. In this case, you need another process through another sophisticated computer algorithm to send your secret keys. Now in the case of two computers, several algorithms are needed. One will encrypt the message (packages) that will be very hard to decrypt by an unauthorized person. Second algorithm will decrypt the message for the recipient. And a third algorithm will transfer your key to the authorized recipients. In all these processes, senders and receivers should not even see any of these elaborate processes and systems that need to be extremely fast, particularly if your packages (documents, text messages, pictures, etc.) are long (Figure 2.1).

Encryption is when data is encoded using encryption keys, which are part of specific algorithms [4]. The encoded data looks like a meaningless language. This encoded data is readable by a person having the decryption key who can decrypt the data and make it readable. Encryption allows confidential information of users, such as passwords, payment details, etc., to remain safe from hackers.

All encryption technologies use two types of key algorithms to encrypt their data: symmetric-key and asymmetric-key algorithms:

- In a symmetric-key algorithm, both the encryption and decryption of data are done using the same key.
- In an asymmetric-key algorithm, both encryption and decryption are done using two different types of keys.

There are different types of cryptographic methods. The difference is mainly related to factors like response time, bandwidth, computational cost, etc. Thus, each of the cryptographic algorithms has its own strength and limitation [5].

FIGURE 2.1 Encryption.

- A symmetric-key algorithm is fast, but there are problems with how to transmit the key safely.
- An asymmetric-key algorithm is slow, but there is no need to transmit keys.

2.4   ENCRYPTION TECHNOLOGIES

Encryptions protect the user from security menaces, illegal capturing of data, unauthorized access, various malicious attacks, etc. Various technologies are used to encrypt the data of the user, such as RSA (Rivest–Shamir–Adleman), Triple Data Encryption Algorithm (TDEA or Triple DEA), DES, Blowfish (cipher), Twofish (cipher), and more. Figure 2.2 shows the taxonomy of main encryption technologies.

2.4.1   RSA (Rivest–Shamir–Adleman)

RSA is an encryption system developed by Ronald Rivest, Adi Shamir, and Leonard Adleman in 1977 (RSA stands for the first letter in each of its inventors' last names). RSA is used to encrypt and decrypt messages. It is an asymmetric cryptographic algorithm. There are two different keys. One of the keys can be given to anyone, and the other key must be kept private [6].

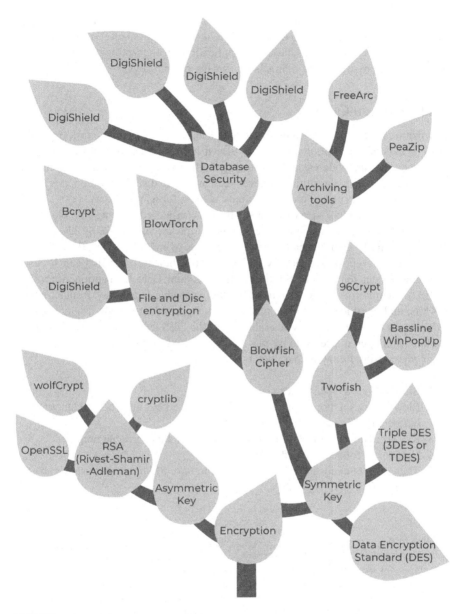

FIGURE 2.2 Taxonomy of encryption technologies.

Once a message has been encrypted with the RSA algorithm with the public key, only the private key can decrypt it.

RSA is slow, so it is seldom used to encrypt data; more frequently, it is used to encrypt and pass around symmetric keys, which can deal with encryption faster.

Public key encryption schemes differ from symmetric-key encryption, where both the encryption and decryption processes use the same private key. These differences make public-key encryption techniques, like RSA, useful for communicating in situations where there has been no opportunity to distribute keys beforehand safely.

Symmetric-key algorithms have their applications in encrypting data for personal use or when there are secure channels for sharing private keys.

2.4.2 BLOWFISH CIPHER

Blowfish was designed in 1993 by Bruce Schneier (schneier.com) as a fast, free alternative to existing encryption algorithms. Blowfish is a symmetric-key algorithm and is very popular and widely used. It is among the first open-sourced encryption algorithms. The Blowfish algorithm has never been broken to date. In addition, it is a very fast encryption algorithm – one of the many reasons for its popularity.

Blowfish uses a block size of 64 bits, which is considered to be a small block size in the encryption world since the typical range for the key size is usually 32–448 bits. Blowfish is highly secure because it uses 16 rounds of encryptions [7, 8].

Blowfish encryption technology is a symmetric block cipher that can be used as a replacement for DES or Simplified International Data Encryption Algorithm (IDEA), a symmetric-key block cipher. It takes a variable-length key, from 32 bits to 448 bits, making it ideal for both domestic and exportable use. Blowfish technology is developed to overcome the vulnerabilities of the DES technique. The algorithm is available and open to use as no patents have been filed for this algorithm. The algorithm works using following parameters:

1. Blowfish has a 64-bit block size.
2. Blowfish uses a large number of subkeys. These keys must be precomputed before any data encryption or decryption.
3. Key size can vary in range between 32 and 448 bits (variable range).
4. Number of rounds allowed is 16.
5. There are four substitution boxes, each can have 512 entries and each entry maybe of 32 bits.

Block Size	Algorithm
A block size of 32-, 56-, or 64-bit encryption refers to key size. Fifty-six bits, or seven bytes, for symmetric encryption is stronger than 40-bit encryption	An algorithm is a sequence of well-defined, computer-implementable instructions to perform a computation. Algorithms are always unambiguous and are used as specifications for performing calculations, data processing, automated reasoning, and other tasks

2.4.3 DATA ENCRYPTION STANDARD (DES)

The DES is a symmetric-key encryption system. Although it's short key length of 56 bits makes it too insecure for modern applications, it has been highly influential in the advancement of cryptography.

Developed in the early 1970s at IBM and based on an earlier design by Horst Feistel, the algorithm was submitted to the National Bureau of Standards (NBS) following the agency's invitation to propose a candidate for the protection of sensitive, unclassified electronic government data. In 1976, after consultation with the National Security Agency (NSA), the NBS selected a slightly modified version (strengthened against differential cryptanalysis, but weakened against brute-force attacks), which was published as an official Federal Information Processing Standard (FIPS) for the United States in 1977.

2.4.4 TRIPLE DES (3DES OR TDES)

Triple DES (3DES or TDES) is a symmetric-key encryption system. TDES applies the DES cipher algorithm three times to each data block. The DES's 56-bit key is no longer considered adequate in the face of modern crypto-analytic techniques and supercomputing power. However, this adapted version of DES, i.e., TDES or 3DES, uses the same algorithm to produce more secure encryption.

2.4.5 TWOFISH

Twofish is a symmetric-key encryption system. Twofish's distinctive features are the use of precomputed key-dependent S-boxes and a relatively complex key schedule.

Twofish borrows some elements from other designs. Like DES, Twofish has a Feistel structure which consists of multiple rounds of processing plaintext, each round consisting of a "substitution" step, followed by a permutation step. Twofish employs a maximum distance separable (MDS) matrix. MDS is a matrix representing a function with certain diffusion properties that have useful applications in cryptography.

FEISTEL STRUCTURE

Feistel cipher is a symmetric structure named after the German-born physicist and cryptographer Horst Feistel, and known as a Feistel network. A large proportion of block ciphers use the scheme, including the DES. The Feistel structure has the advantage that encryption and decryption operations are very similar, even identical in some cases, requiring only a reversal of the key schedule. Therefore, the size of the code or circuitry required to implement such a cipher is nearly halved.

2.5 ADVANTAGES AND DISADVANTAGES OF BLOWFISH AND RSA ENCRYPTION

2.5.1 ADVANTAGES

2.5.1.1 RSA

Some advantages of the RSA algorithm are that it is safe and secure for its users through the use of complex mathematics. Additionally, it's difficult to crack. Another advantage is that the algorithm uses a public key to encrypt data and the key is known to everyone. Finally, RSA encryption is good for protecting the transfer of data.

2.5.1.2 Blowfish

Blowfish is one of the fastest ciphers in general use. A preprocessing is required for each key, and it is equivalent to encrypting approximately 4 kilobytes of text – slow as compared to other block ciphers. Thus, it is used in certain applications such as SplashID (a password management application) but not others. The password hashing method used in OpenBSD (a security-focused, free and open source, Unix-like operating system) is a benefit because it uses an algorithm that makes use of the slow key schedule derived from Blowfish. It is freely available as it is not subject to any patents and is thus open to public use. This was the major reason contributing to the popularity of cryptographic software.

2.5.2 DISADVANTAGES

2.5.2.1 RSA

The RSA algorithm can be very slow in cases where large data needs to be encrypted by the same computer. For example, if one wants to send a sensitive file, encrypting it with RSA is going to be difficult due to the low performance of RSA encryption.

Also, RSA requires a third party to verify the reliability of public keys. Data transferred through the RSA algorithm can be compromised through middlemen who might temper with the public key system.

2.5.2.2 Blowfish

Blowfish gets the key to the person through an unsecured transmission channel. Key management is also a big problem as every user needs a unique key; so as the number of users increases, the key management becomes problematic. Since two people have the same key, Blowfish can't provide authentication and non-repudiation (protection against denial). Also, this algorithm has high time consumption, and its serial throughput is high because of its weakness in the decryption process.

2.6 WHICH PRODUCTS USE ENCRYPTIONS

A standard algorithm should apply to different products. Some applications of an algorithm include the following:

- Bulk encryption: an algorithm should be able to encrypt a large number of files or continuous streams of data.
- Random bit generation: the Blowfish algorithm should be capable of generating nine block cipher Blowfish single random bits.
- Packet encryption: the algorithm being used should efficiently encrypt packets of data. Any application that requires packets to be encrypted or decrypted should be able to implement an algorithm.

2.6.1 PRODUCTS USING BLOWFISH

Some examples of products using the Blowfish algorithm include:

- File and disc encryption (Bcrypt, BlowTorch, DigiShield)
- Password management (1Password, KeyRing Mini, Password Wallet)
- Archiving tools (FreeArc, PeaZip – Backup: Leobackup, Crashplan)
- Database security (Encryptionizer, SQL Server 2000 Encryption)
- Operating systems (Linux, Novell SUSE Linux Enterprise)

2.6.2 PRODUCTS USING RSA

RSA encryption is good when there are two physically or geographically different endpoints. RSA encryption is good for protecting the transfer of data across geographic boundaries. The RSA algorithm is often used for securing communications between web browsers and eCommerce sites.

2.7 CONCLUSION

Blowfish is one of the fastest algorithms for data encryption. It uses a block cipher, which makes the encryption and decryption process quite secure and reliable. The Blowfish algorithm is suitable for all applications where the key is not getting changed frequently, as the communication link. The algorithm works faster as compared to its predecessor DES. It uses the 16-pass block encryption algorithm which increases security. Also, the encryption and decryption processes cannot be done on very large-size files (exceeding 4GB). This algorithm has been tested under various conditions and in each condition, it has been found to be the most secure and unbreakable algorithm. The built-in instructions which are present in the microprocessors for the shifting of the bit operations are quite advantageous for the Blowfish algorithm, making it faster. The algorithm is compatible with most hardware, making it quite efficient to use. Figure 2.2 summarizes the main taxonomy of encryption technologies.

RSA encryption is usually used when you have two physically separate endpoints. It is often used in web browsers to connect to your favorite websites, in VPN connections, and in many other applications. As such, we are all using asymmetric encryption every day, possibly unbeknown to us.

RSA differs from older symmetric algorithms because it has the ability to allow digital signing as well as advanced encryption for online commerce systems and high-profile institutions. RSA is widely used to secure financial, personal, and corporate-level information for a variety of domestic and commercial users. It is used in various applications such as Bluetooth, MasterCard, VISA, e-banking, and e-communication. Figure 2.2 shows taxonomy of encryption technologies.

FACEBOOK STORED "HUNDREDS OF MILLIONS" OF UNENCRYPTED PASSWORDS IN PLAINTEXT

In 2019, Facebook admitted that it mistakenly stored "hundreds of millions" of passwords in plaintext, unprotected without any encryption.

Facebook kept user passwords in its internal servers in an insecure way, affecting hundreds of millions of Facebook Lite (an Android app designed for low-speed connections and low-spec phones) users, tens of millions of other Facebook users, and tens of thousands of Instagram users.

According to security reporter Brian Krebs, who cited a "senior Facebook insider," "access logs showed some 2,000 engineers or developers made approximately nine million internal queries for data elements that contained plaintext user passwords," as reported by *The Guardian* [9].

TWITTER BUG SECURITY FLAW EXPOSED PASSWORDS IN PLAINTEXT

In 2018, Twitter advised all 330 million of its users to update their passwords after a bug in the password hashing process resulted in passwords being saved in plaintext within an internal log.

Twitter conducted an investigation and found no evidence of a breach or misuse of these passwords, but still responded and publicly urged all users to change their passwords.

According to Archie Agarwal, the CEO of the cybersecurity company ThreatModeler, "If all the 330 million passwords were stored in clear text in an internal log, then it's not really a bug but a design flaw," as reported by CNET [10, 11].

3 Authentication

3.1 INTRODUCTION

Authentication is a technique or mechanism to prove and validate an end user or a computer program's identity. People used to authenticate based on voice, a word or short, written documents, etc. In recent times, personal identification numbers (PINs), passports, driving licenses, and other IDs are used.

Computer authentication was created in the early 1960s to have an access control mechanism to share the limited computing resources of expensive mainframe computers. Authentication managed access control of the system by only allowing the users with access privileges.

To overcome the problems with legacy authentication methods, more sophisticated technologies were gradually invented, such as digital certificates, biometrics, access tokens, security keys such as Yubikeys (a USB device that generates a unique passcode) and Zukey (security token hardware). Today, a combination of PINs and security questions, passwords, etc. are used.

This chapter will explain authentication technologies, their advantages and disadvantages, and the different products, which use these technologies.

3.2 BRIEF HISTORICAL BACKGROUND

In the early 1960s, computers were expensive, bulky, and relatively slow, and there was usually only one computer available to share amongst a large group of people. Due to the large demand for this single resource, universities like Massachusetts Institute of Technology (MIT) developed time-sharing operating systems such as the Compatible Time-Sharing System (CTSS). To keep files private for the respective users, in 1961, Fernando Corbató, an MIT researcher, used passwords to protect user files on the multiuser time-sharing system. Later, another PhD researcher, Allan Scherr, involuntarily discovered the weakness of a password system. Passwords needed to be stored somewhere to validate them against a user's input. To increase usage time on the CTSS, Scherr found the password file location and printed it, gaining access to all the passwords in that system. The CTSS password leak demonstrated that storing passwords in a clear text file was not a good idea. Learning from this incident in early 1970s, a Bell Labs researcher, Robert Morris, developed a way to obscure the passwords in the master password file by using a cryptographic concept called a hash function. A hash function is any function that can be used to map data of an arbitrary size to fixed-size values. The values then are used to index a fixed-size table called a hash table. With this technique, it was possible to verify the passwords without storing them as clear text.

DOI: 10.1201/9781003038429-3

During the 1980s, the industry was persistently looking for highly secure and dynamic methods of authentication, as many researchers and hackers were finding new ways to abuse the persistent password system. Thus, a new concept called one-time passwords (OTP) wherein the user had a new password every time they logged in was created. The initial implementations of the OTP system required the user to have special hardware to receive the password. That said, some other systems were also able to provide OTPs via more accessible channels such as text messages, the web, etc.

In early 2000 the Multi-Factor Authentication (MFA) appeared, which was a combination of two or more authentication factors to identify an entity more securely. The bare minimum MFA is the Two-Factor Authentication (2FA), which uses just two authentication factors. MFA solutions can technically have as many authentication factors as the user wants, for example, combining a password with an OTP [1].

3.3   HOW AUTHENTICATION TECHNOLOGIES WORK

ANALOGY

To better understand authentication, it could be helpful to think of it as immigration and customs. Today when we travel to a foreign country, we must have a passport recognized by the state that we enter. The passport needs to be valid, and there are immigration officers who inspect the passport to ensure it is not forged or falsified. This process is similar to the various elements of authentication that we discuss (Figure 3.1).

3.3.1   SECRET KNOWLEDGE-BASED AUTHENTICATION

In knowledge-based authentication, the user has to memorize a secret (sequence of inputs), which consists of either just numbers (PIN), answers to predefined questions (cognitive knowledge), or images (graphical password) [2]. The secret is known by both the user and the system, which means there needs to be an exact match between them for successful authentication. This means that it is a zero/one authentication process, meaning the outcome is either true (a completely true secret resulting in access being granted) or false (totally false secret resulting in the denial of access). This means that there is a reliance on the user's ability to recall the exact secret when asked, irrespective of the length or complexity of the secret.

3.3.1.1   Personal Identification Number (PIN) and Password

A PIN is considered the easiest method for authentication. It can be used within phones to lock/unlock and with debit/credit cards. Typically, a mobile PIN ranges from 4 to 8 digits. Since numbers are relatively simpler to recall, guessing and stealing them is also easier.

Passwords, which can be longer and are made of numbers, letters, and symbols, lessen the possibility of being cracked. They are believed to offer effective protection if they are established and employed appropriately.

FIGURE 3.1 Authentication.

3.3.1.2   Cognitive Knowledge Question

Cognitive knowledge, which is in the form of questions, aims to lessen the load of users memorizing complex passwords, thereby utilizing associative questions. These questions are usually about personal information (i.e., mother's maiden name, first teacher, favorite pet, etc.) or preferences, such as a favorite holiday destination or a favorite movie. As such, this method lacks one of the most important characteristics of a secret knowledge-based authentication approach, which is secrecy. By guessing or researching online, it is possible to find the answers to these kinds of questions, thereby making them vulnerable. Hence, this method cannot be used on its own, and instead a group of CKQs should be used. Even though this solution does probably increase the security by adding an additional layer, it also increases the burden on the user to remember answers, thereby increasing the authentication time and requiring the user to recall multiple secrets [2].

3.3.1.3   Pattern and Graphical Password

Many of the solutions that have been suggested to lessen the drawbacks of using PIN, passwords, and passphrases do not address the human inability to retain and recall multiple complex passwords.

The human mind is more capable of storing and remembering visual information as opposed to textual information [3]. Because of this, pattern password authentication has emerged, in which a user is most commonly required to draw a preset outline on a 3 × 3 dot grid that appears on a touch screen. Therefore, it is suggested that it will be much easier and more convenient for the user to recognize a visual pattern as opposed to an alphanumeric password. Mobile devices with touch screens make it very convenient to utilize pattern passwords – which are predominantly used in Android devices to improve the memorability of the secret.

3.3.2   TOKEN-BASED APPROACH

Tokens have been developed to overcome the drawbacks of the secret knowledge-based approach. Tokens can be classified into two categories based on their applications: hardware tokens and software tokens [4]. In hardware token systems, a separate physical device is produced and provided, usually by the service provider, such as a bank smartcard or HSBC secure key OTP token [5]. In software token systems, an existing device is used, such as sending the OTP to the user's registered mobile phone.

3.3.3   BIOMETRICS

Biometrics is a technology that is based upon measurable and distinctive characteristics of a user. They can be categorized based upon their characteristics into physiological and behavioral biometrics.

Physiological biometrics are based upon certain unique physical aspects of the body, such as the face, a fingerprint, or the iris/retina in the eye. Behavioral biometrics use the distinctive way in which users behave, such as keystrokes, voice, and signatures, to identify and authenticate a user. Both categories uniquely identify individuals. The secret is non-transferable to others, unforgettable, cannot be easily lent or stolen, and difficult to reproduce, change, or hide. However, the error rates and cost of biometric systems, along with their usability, have hindered their widespread adoption. Also, biometric systems can be fooled and cracked, although with some difficulty. For instance, facial recognition can be fooled by a photo of the authorized person. Similarly, voice recognition can be fooled by voice imitation or voice recordings. Biometric systems have also been reported to be racially biased [6].

Biometrics could be used in combination with a token, password, or additional data in order to achieve more security. Techniques have been suggested and implemented to determine whether the provided biometrics sample is from a living and legitimate user. An example of this is Liveness detection (used to detect a spoof attempt by determining whether the source of a biometric sample is a live human being or a fake representation) utilizing some biological indicators, such as temperature and pulse for fingerprint systems and blood flow and blinking for an iris scan [7] (see Chapter 4).

3.3.4 Compound Authentication

To increase the level of security, two or more authentication techniques can be employed in combination. They can consist of multiple techniques from the same authentication approach (multi-layer authentication), such as password and cognitive questions, or from different authentication approaches (multi-factor authentication), such as PIN and smart card, password and facial recognition, or fingerprint and OTP generator token. This can then be reinforced by elements such as a predefined user location, which can be based on either the mobile cellular network (i.e., cell phone ID), the global positioning system (GPS) (i.e., longitude, latitude) [8], or/and the IP address.

3.3.5 Storing Passwords and Usernames

The first step of account creation is to create a username and password with other types of user authentication factors. The authentication factors then need to be stored in a database. It is obvious that such data should never be stored as text and must be encrypted; if a data breach were to happen, hackers could access the authentication factors and see the ciphertext in the password field. Consequently, the fundamental step in safeguarding authentication factors is the encryption of these factors. However, encryption on its own is not enough to secure authentication factors. In this case, other technologies are needed such as hashing passwords and salting.

Hashing Password: Hashing converts passwords into unreadable strings of characters that are designed to be impossible to convert back, called the hashed password.

Salting: Salting adds a unique value to the end of the password to create a different hash value. The additional value is called a "salt." Salting adds a layer of security to the hashing process, specifically against brute force attacks. A brute force attack is where a computer or botnet (a number of Internet-connected devices which might run one or more bots) attempts every possible combination of letters and numbers until the password is found.

By adding a salt to the end of a password and then hashing it, you've essentially complicated the password cracking process.

3.4   AUTHENTICATION TECHNOLOGIES

Authentication technologies encompass various products and services which implement a wide range of methods to authenticate a user's identity. There are three types of authentication to be considered:

1. Accepting proof of identification given by a trusted person who guarantees that the person is genuine, such as public certificate authorities, a web of trust (decentralized peer-based trust), etc.
2. Authenticate by comparing the attributes of the object itself, for example, identity certificates with hard to duplicate features like fine prints, holographic images, and watermarks.
3. Authenticate based on documentation or affirmations from authorities. In a computer system, valid user credentials can imply authenticity as an authorized network administrator gives it.

To store passwords is generally stored in hashed form. This technique is used to protect passwords from hackers who gain unauthorized access to the database. When a password is hashed, it is turned into a scrambled representation of itself. A user's password is taken and – using a key known to the site – the hash value is derived from the combination of both the password and the key, using a set algorithm.

Hashed passwords with salts (when a cryptographic salt is made up of random bits added to each password instance before its hashing), public-key cryptography, private-key cryptography, OTP, etc., have emerged to authenticate users with their computing systems. But as computers have become increasingly powerful, the traditional methods of authentication have become weaker and more prone to being cracked by brute force methods – when an attacker uses an algorithm that tries all possibilities until a solution to the problem is found.

3.5   ADVANTAGES AND DISADVANTAGES OF AUTHENTICATION TECHNOGLOGIES

3.5.1   ADVANTAGES

3.5.1.1   Personal Identification Number (PIN) and Password Approach

- Simple to use and implement
- Easy to change
- Most common form of authentication technology
- Complex and long passwords are nearly impossible to guess

3.5.1.2   Cognitive Knowledge Question (CKQ)

The strength of cognitive passwords can be measured by the ratio of memorability and guessability. Advantages of CKQ include:

- Appropriate for a majority of the population.
- Answers with high remembrance.
- Questions with a single correct answer.

- Answers cannot be discoverable through research.
- The memorability of cognitive questions is stable over time.
- The recall rates for cognitive questions are higher than that of passwords.

3.5.1.3   Pattern and Graphical Password
- A large variety of pattern options are available.
- Makes guessing by the intruder very difficult [10].

3.5.1.4   Token-Based Passwords
- More secure authentication method than passwords and PINs.
- The token validity is time-based. Some tokens change every 60 seconds, and some are only valid for 15 minutes.
- A token-based password can only be used once. They become invalid after a single-use.

3.5.1.5   Biometric Authentication
Biometric Authentication is of several types:

- Fingerprint Scanning
- Facial Recognition
- Voice Identification
- Retina and Iris Scanning

Below are some of the generic advantages:

- A much-improved level of security.
- No need to worry about forgetting it.
- A much faster way of authenticating; the response time is speedy.
- It cannot be replicated by anyone [7].

3.5.1.6   Compound Authentication
- Strengthens security by providing an extra layer of authentication.
- Each factor in the system is built upon the weakness of the previous factor.
- Reliable against brute force attacks and protects from hacking attempts such as keyloggers.

3.5.2   Disadvantages

3.5.2.1   Personal Identification Number (PIN) and Password Approach
- To remember them easily, users create a common password, which can be guessed easily by hackers.
- The PIN is a weak mode of protection since people might use a very simple and guessable PIN like "1234."
- Intruders can use methods such as blacklisting to decline service to users by purposefully making consecutive unsuccessful login attempts and blacklisting them.
- Passwords, if not properly stored, can be stolen via a dictionary attack [8].

3.5.2.2   Cognitive Knowledge Question

- These questions are more dependent on the memory of a user. Compared to passwords, the guessability rate is high.
- Only certain questions have acceptable ratios, not all.

3.5.2.3   Pattern and Graphical Passwords

- People often don't use a long or confusing pattern, they use a very basic pattern, which makes it easy for an intruder to guess and/or replicate.
- Patterns can be difficult for people to remember for a long time.
- Pattern lock is one of the most insecure methods to use for authentication on smartphones [9].

3.5.2.4   Token-Based One-Time Passwords (OTP)

- Users need an additional device with them to receive an OTP.
- In the case of a hardware token, users are required to carry the token device with them in all cases.

This method may involve additional costs, including the cost of the token device, replacement costs, application, etc.

3.5.2.5   Biometric Authentication

- Biometric systems are not 100% accurate.
- This technology requires integration and additional hardware, which might be expensive.
- The usage and surroundings of the users can affect the accuracy and measurements.
- Biometric systems need to have frequent security updates in order for them to work properly.
- There is no possibility to generate a new one, as password if compromised.

3.5.2.6   Compound Authentication

- The cost of setting up multi-factor authentication is high.
- One of the biggest drawbacks of multi-factor authentication is the account recovery process. This method relies upon additional authentication, and users need to memorize more details [10].

3.6   WHAT PRODUCTS USE AUTHENTICATIONS

In the race for innovation, new products exploit the full potential of security and authentication services serve as their backbone. The Secret Knowledge-Based approach for authentication is employed in almost all web applications as the primary source of authentication for its users. They are primarily used in online applications since they are zero/one based and have to rely on the exact secret key match by the user. Most web interfaces and products use this kind of approach.

The PIN-based authentication approach can be seen mostly with banking and financial applications, ranging from chipset cards to ATM machines. The PIN-based service model is used at every level and is more common in these applications because of the fixed length of alphanumeric characters, which make it easier to propagate to large-scale masses.

Cognitive knowledge-based authentication is an extension service in authentication and cannot serve as a standalone mode. These kinds of services are mostly compounded to the secret-based approach to add an additional level of check. They are used in web portals as an integral part of the Sign-Up module or Password reset module, which requires authentication of the user at a more personal level.

Handheld devices use almost all authentication methods available. However, Android-based devices employ a pattern-based authentication to put a lower cognitive load on their consumers for quick service. Most mobile devices employ this technique, and it has proven to be an effective and popular medium of authentication.

Enterprise solutions and products use a token-based approach. Employees either use a hard token on a small handheld device or a soft token on their mobile devices that generates a random key which is always in sync with a private server to match up. This is a more technical and expensive solution; hence, mostly enterprises make full use of this.

Apart from these techniques, biometric solutions that require the physical presence of the person to authenticate are mostly used to secure workplaces or restricted areas. Products such as biometric cameras, scanners, and fingerprint readers make the most use of biometrics. Furthermore, laptops use this as well in the form of scanners and readers too.

Additionally, apart from using all these techniques individually, often a combination of these techniques are used. Some banking portals and mobile applications use an additional layer of security. The army uses a cluster of authentication techniques to make sure important assets are handled safely.

3.7 CONCLUSION

Authentication services have played a vital role in safeguarding users with their important assets time and again. There has been a massive improvement in the overall framework for these services, and the last decade has shown the introduction of several new products.

Many authentication techniques have become obsolete, and many have been adopted slowly and gradually on a large scale.

Since the length of alphanumeric characters is limited in a PIN-based approach, it is exposed to increased vulnerability, especially with certain hacking algorithms. That said, banking services often use this method on a very large scale since they have an effective infrastructure to support this type of authentication.

The cognitive knowledge approach is not very popular and effective if used alone and requires human effort and time to deal with the mechanism. It has been found to only be effective if combined effectively with other approaches.

The pattern password-based approach has found its fair share of users in mobile device consumers and has been very effective thus far in providing easy and quick solutions. However, pattern password-based authentication is limited in its reach since it is confined to mobile devices only, as not every product can support such a security pattern.

The token-based approach has been very effective and secure for authentication, but it comes with the cost of setting up the dedicated resources for token generation, handling, and matching. Therefore, if cost is not a large factor, this has proven to be a very effective and secure solution. Figure 3.2 summarizes the main taxonomy of authentication technologies.

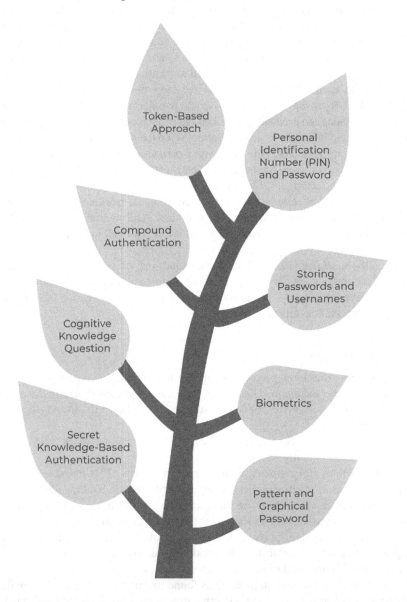

FIGURE 3.2 Taxonomy of authentication technologies.

Biometrics have been very effective solutions to deal with the restricted access to a workplace and are one of their kind in their domain. As such, no other solution comes close in this domain, and they are the sole winners when it comes to a physical presence requirement.

Lastly, a multi-layer and multi-factor authentication is the most useful approach for this innovative era where hackers have been just as creative and innovative as the solutions they try to attack. A PIN-based approach, followed by facial recognition and complemented by GPS latitude and longitude with IP address tracking, can be a very effective solution.

2019 CAPITAL ONE CYBER INCIDENT

On March 22 and 23, 2019, Capital One announced that a hacker broke into a server by exploiting a configuration vulnerability in a web application firewall. Paige Thompson, a 33-year-old Seattle resident, accessed personal information for more than 100 million Capital One customers in the United States and 6 million in Canada.

"The largest category of information accessed was information on consumers and small businesses as of the time they applied for one of our credit card products from 2005 through early 2019. This information included personal information Capital One routinely collects at the time it receives credit card applications, including names, addresses, zip codes/postal codes, phone numbers, email addresses, dates of birth, and self-reported income" [11, 12].

4 Biometrics

4.1 INTRODUCTION

Biometrics provide good security since every individual's physical and behavioral traits are permanent, unique, and non-replicable. The use of biometrics is now common throughout the world for identification of documents, mobile devices, banking, and other functions.

This chapter reviews biometric technologies, explains how the technologies work, their advantages and disadvantages, and the different products which use biometrics.

4.2 BRIEF HISTORICAL BACKGROUND

Biometrics were first used in the 18th century by Alphonse Bertillon, a French anthropologist who used anthropometrics (body physical measurements) to identify criminals. The use of fingerprints is actually much older, as they have been found on ancient Babylonian clay tablets, seals, and pottery, on the walls of Egyptian tombs, in the Minoan and Greek civilizations, and in Ancient China where they authenticated government documents with fingerprints.

In the 1870s, while working in colonial India, William Herschel also recognized the unique qualities of fingerprints and began using them as a form of a signature on contracts with locals. Edward Henry developed and first implemented the system in 1897 in India, named the "Henry Classification system." In 1901 the Henry system was introduced in England; shortly after, the New York Civil service began testing the Henry method of fingerprinting and by 1907, the US Army, Navy, and Marines had all adopted the technique.

The 1960s brought the introduction of iris pattern recognition, facial recognition, fingerprint automation, automated signature recognition, face recognition automation.

From the 1970s to now, biometric technologies have developed and improved significantly, and this led to the 2002 creation of ISO/IEC standards committee on biometrics.

In recent years, using biometrics has expanded in many areas and is still growing. For example, in 2008, the US Government began coordinating a biometric database. In 2010, the US national security apparatus utilized biometrics for terrorist identification. In 2013, Apple included fingerprint scanners in their smartphones. Since then, the usage of biometrics for authentication, surveillance, and identification only continues to increase.

DOI: 10.1201/9781003038429-4

4.3 HOW BIOMETRIC TECHNOLOGIES WORK

ANALOGY

The use of biometrics can be analogous to something we've seen in science fiction movies for years: instead of using a card or entering a PIN or password to open a door, a system uses your eye, fingerprint, face, or maybe even your brain waves, DNA, or other biological measurements of your body to authenticate you (Figure 4.1).

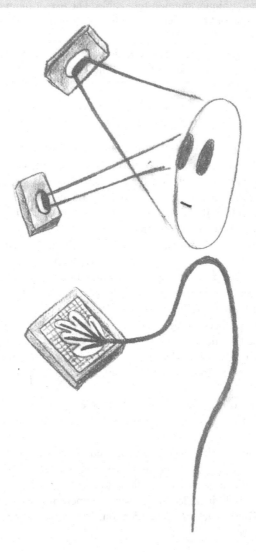

FIGURE 4.1 Biometric.

Biometrics utilize physical or physiological traits that make each human unique rather than using something we have (like a key) or something we know (like a secret phrase). Biometrics utilize physical qualities, similar to our face, fingerprints, irises or veins, or consider attributes like our voice, or penmanship [1, 2].

In contrast to keys and passwords, our characteristics are hard to lose or overlook. For that reason, they can be exceptionally difficult to duplicate. Thus, it is believed that they can be more secure than keys or passwords.

Biometric frameworks can appear to be confusing, yet they are all utilized similarly in three stages:

Enrollment: The first time the user utilizes a biometric framework, it records essential data about the user, similar to a username or a recognizable proof number. It, at that point, catches a picture or recording of a particular user attribute.

Storage: Most biometric frameworks don't store the total picture or recording. They instead break down user characteristics and interpret them into a code or graph.

Comparison: Whenever one utilizes the framework, it looks at the attribute user present to the data on the document. At that point, it either acknowledges or dismisses that the user is who they claim to be.

Biometric frameworks additionally utilize three things:

- A sensor that identifies the trademark being utilized for ID.
- A computer that stores and reads the data.
- A software that examines the characteristics and interprets them into a graph or code and carries out the comparison.

4.3.1 PALM

The human palm contains characteristics that can be used for identification and accurately identify people of all races and skin tones. For example, palm vascular can be used by comparing the pattern of veins in the palm of the person being authenticated with the pattern stored in a database. Since vascular patterns exist inside the body, they make it extremely difficult to forge or replicate. The palm normally does not have hair and is also less susceptible to a change in skin color, which makes it ideal for authentication.

4.3.2 FINGERPRINTS

The first step is to obtain a user's fingerprints with a scanner. The scanner records information of the fingerprints and stores them in the form of a code. The code can later be used to verify if the fingerprints are the same.

There are three types of scanners:

4.3.2.1 Optical Scanners

Optical scanners use LEDs to take a picture of the fingerprint. They use algorithms to find patterns on the surface of the scanners. Patterns include unique marks and ridges on the fingerprint. The scanner uses the light that reflects the fingerprint and determines the lightest and darkest parts of the picture to capture this information. The main drawback is that a 2D image of the fingerprint can easily replicate or fool a system like this.

4.3.2.2 Capacitive Scanners

These are the most commonly used scanners and are often found on many mobile phones nowadays. They use many small capacitor circuits that can store electricity. Because of the ridges in a fingerprint, a different amount of charge is stored in the capacitor, and this information is stored for future comparison. These are more secure than optical scanners as they cannot be fooled by an image.

4.3.2.3 Ultrasonic Scanners

Ultrasonic scanners use ultrasonic pulse to capture information about a fingerprint. The pulse bounces back from the fingerprint ridges, some of which are absorbed due to ridges and other unique characteristics of the fingerprint. A sensor detects how much signal is being bounced back and as such, a highly detailed three-dimensional image of the fingerprint can be captured. As a result, ultrasonic scanners can be more secure than capacitive scanners.

4.3.3 FACE

Facial recognition technology works by capturing the features of a face and stores the information for later comparison. The human face has many different marks, highs, and lows that differ from individual to individual, such as the distance between the eyes, width and length of the nose, the depth of the eye sockets, and the length of the cheekbone jawline, etc. This information is used to create a faceprint.

4.3.4 IRIS

The iris is another unique part of the human body that does not change over time. A digital camera is used to scan the patterns that are present on the iris of the eyes. The camera locates different parts such as the center and the edges of the pupil, eyelids, center of the iris and eyelashes. Then this information is converted into a code that is stored.

4.3.5 VOICE

Voice is also unique to every individual. Voice biometrics digitize the voice of an individual and produces a voiceprint or a template. Voice systems either use a specific phrase or a sentence or enough voice input is provided to the computer algorithms, which are used to identify the voice. The voiceprint of an individual is then stored, similar to other biometric technologies.

4.4   BIOMETRIC TECHNOLOGIES

Biometrics are used in cybersecurity for authentication and verification. In general, biometric systems have various uses including: monitoring, law enforcement, time and attendance, logical access controls, and control of physical access. Other applications of biometric systems include [3]:

- Securing surveillance systems
- Physical access to buildings
- Applications such as authentication for computer network, online banking, e-commerce website, border control, medical records management, and security monitoring
- Providing confidence in the verification and authentication of users
- Identifying individuals and monitoring access to information, physical spaces, and services
- Enhance health, reduce fraud, and reduce national security threats

A biometric authentication system or identity verification (IV) system is used for authenticating users on the web, mobile, or any digital platform by using the biometric features of the individuals. The feature characteristics used for biometric authentication can be broadly classified into two classes (also shown in Figure 4.2):

- Physiological characteristics related to the shape and size of a person's body. These features are natural, unique and differ from one person to another. Examples include fingerprint matching, facial recognition, and iris matching.
- Behavioral characteristics related to the habits, practices, and behavior of the person. Examples include the signature, keystroke patterns, and voice of the person.

The biometric authentication ecosystem consists of a sequence of six processes. The process begins with inputting the biometric data of the individual into the system. The data is then preprocessed to find the areas of interest from which features can be extracted. Once the areas of interest are identified, the next step in the process is to implement extraction algorithms using isolated or hybrid classification techniques to extract unique and non-replicable features of the individual. Finally, the feature matching and decision modules are implemented using deep learning techniques.

4.5   BIOMETRIC TRAITS COMPARISON

Each biometric trait has multiple features. The following summarizes the main biometric trait features:

Fingerprint

- Accurate and reliable technique
- Extremely fast and easy to process, store and match
- Requires the user to come in direct physical contact with the system

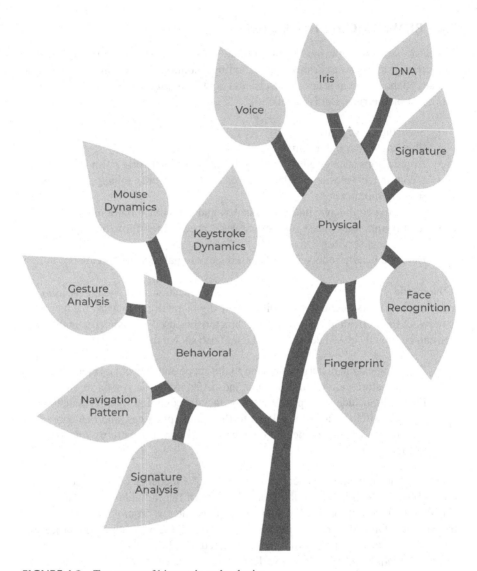

FIGURE 4.2 Taxonomy of biometric technologies.

Face

- Relatively new technique
- Very efficient algorithms required as people can have similar facial traits
- Involves no physical contact

Iris

- Very efficient technique
- Not very cost-effective
- Requires a user to be in close proximity to the system

Voice

- Very cost-effective
- Easy implementation
- Throat infections can lead to inaccuracy and inability of legitimate users to access the system

Signature

- Widely used
- Very easy to implement
- Highly reliable

Types of Biometrics and their applications

Fingerprint

- Authentication
- Law enforcement
- Identity verification
- Access control

Face

- User verification
- Human-computer interaction
- Surveillance
- Access control
- Criminal identification
- User registration

Retina

- Used in highly secured agencies for authentication and verification of humans

Iris

- Mobile phone access
- National security on land, sea, and airports

Palm Print

- Student authentication systems
- Blood relation identification

- Medical diagnosis
- Personal identification

Voice

- Voice response (health and banking systems)
- Web transactions

Signature

- Legally used for authentication

4.6 ADVANTAGES AND DISADVANTAGES OF BIOMETRIC TECHNOLOGIES

While biometric systems have great advantages over traditional systems such as ID cards and passwords, biometric data are subject to numerous attacks including direct and indirect attacks. Direct attacks are system attacks without machine knowledge. Indirect attacks involve knowledge of the biometric system's internal workings. Here are the types of attacks performed on the biometric systems:

- Fake Biometric: Used to trick the machine and show false results. Fake values are fed to the sensor.
- Resubmitting Previously Used Biometrics: The biometric signal is prerecorded in the device and prestored for several attacks.
- Overriding: The original set of functions is corrupted by an attacker and replaced with a false one.
- Compromising Stored Templates: An attacker accesses and steals certain templates to fake the user's identity.
- Overriding the Decision: Spoofer overrides the final decision.
- Trait Aging: Biological changes in a human over time result in a change of the biometric value, and the transition is inevitable. Changes in the appearance and facial structure of an individual are examples of biological aging, which affect the levels of accuracy.

The biometric system comes with various privacy and security issues. Before using biometric methods, it is essential to address the following problems and concerns:

- Biometric technology is open and an available technology for attackers.
- Biometric traits can be easily trapped and stored by attackers.
- Biometric data cannot be changed; it is always unique.
- Voice recording and surveillance cameras can be used to obtain and replicate the user's voice and facial characteristics.

4.6.1 ADVANTAGES OF USING BIOMETRICS

The advantages of using a biometric system are:

- It provides security by generating unique biometric measurements for every individual and thus generating secure systems. These metrics are closely linked to an individual so no one else can generate the same biometrics.
- It can be efficiently used in the digital world for security and authentication.
- It is currently being used in high-security applications like banking, forensics, defense, surveillance, and personal identification for authentication and identification.
- It does not require an individual to remember any password for identification or authentication; it is something that an individual has in him/herself.
- It can have multiple features that can be grouped to form a highly secure system to make it free from spoofing.
- It provides faster access to applications than traditional systems by quickly authenticating the user. Example – banking applications on phones are easier to access with fingerprint biometrics than with password authentication.
- It is subject to less financial fraud and security threats.
- It is proven to be more reliable for user identification than traditional systems like passwords, pins, smart cards, keys, etc.

4.6.2 DISADVANTAGES OF USING BIOMETRICS

The disadvantages of using a biometric system are:

- Biometric data is sensitive and, if stolen, may result in easy access to the system.
- Biometric data might be subject to direct and indirect attacks.
- It is weak communication channel.
- The algorithms are not suitable enough to extract features for biometrics.
- It can have inefficient results if there are circumstances of dirt, oil, or other material on the physical surface of the system.

Aside from the general challenges listed above, there are additional disadvantages for each of the types of biometrics, as listed below.

Face biometrics

- Data privacy concerns with facial recognition
- High implementation costs
- Data storage
- Changes in appearance and camera angles

Iris biometrics

- Systems require an IR light source and sensor.
- Retinal scanners require a person to look at the scanner from a few inches and thus can't be used in banking and other systems.

Palm print scanners

- Usually bulky and more expensive since they need to capture a larger area than fingerprint scanners.

Voice recognition biometrics

- Not as accurate as other biometrics.
- Requires the detection of liveness to verify that a sample is from a live speaker and not a recording.
- Background noise can impact the quality of the sample.
- Not ideal for all environments.

Signature testing

- Intended to validate subjects based on their special signature characteristics.
- Can be difficult for individuals who do not sign their names regularly to enroll in a signature verification.

4.7 WHAT PRODUCTS USE BIOMETRICS

4.7.1 CARS

Many cars these days have some sort of biometric technology. The ability to use voice recognition to enable mapping and braking systems or to navigate a vehicle is nothing new; it complements a wide variety of features that support health, fuel efficiency, and improves the driver experience in vehicles [4].

4.7.2 BANKING

As international financial institutions become more digitally based, banks are adopting biometric technologies to improve the management of customer and employee identity in an effort to fight fraud, increase the security of transactions, and improve customer convenience. Even consumers are fed up with identity fraud and the inconveniences associated with having to continually prove their identity. Because more and more customers are finding banks that use biometric authentication, banks increasingly study biometric technologies more closely.

4.7.3 HEALTH CARE

Biometrics can benefit various aspects of the health-care system. Biometric data, for example, might allow speedier identification of patients in emergency situations. It could also help deter medication abuse and incorrect orders, which are common problems in the health-care sector. Biometrics could also ensure the privacy of patients, allowing access to their health details only to those who have permission. Licensed doctors may use a simple iris or fingerprint scan to check their identity. The use of facial recognition or fingerprints as part of multi-factor authentication may also advance telemedicine through accurate patient identification and safe digital access to the attending doctor's patient details.

4.7.4 FOOD AND BEVERAGES

Biometrics help to make the food and beverage industry safer and more effective. Food and beverage manufacturing plants are commonly located across several locations around the globe. Biometric technology could allow them to track access rates and permissions of employees globally. This reduces the risk of boundary-contamination, as various rates of access can be required for different staff members, thereby limiting employees to reach some production lines. In addition, biometric systems could help stop any unauthorized people from entering their facilities. For example, Coca-Cola uses a biometric fingerprint system to monitor the behavior of independent truck drivers who come to certain canning site.

4.7.5 BORDER CONTROL

According to the Automated Biometrics Identification System (ABIS), scanned fingerprints help to easily identify someone with a felony track record. Fingerprints, however, have a high degree of incorrect acceptance and rejection. Several countries have been studying iris scans and facial recognition as more effective ways of recognizing passengers, including Thailand, the United Kingdom, Canada, and the United States. With the growing prevalence of self-serve kiosks at airports, departments of government security are gradually gathering a database of eye, iris, and fingerprint scans to help identify potential terrorists or criminals [3, 5].

4.7.6 EDUCATION

Everything that needs verification, from the lunch program to the dorm, can be done with face recognition or fingerprints. School professors can use a similar model to access grades and personal details for students. School protection has recently become a growing issue in the United States. Any unauthorized activity inside schools can be easily detected through facial recognition. Academic integrity could be better preserved using artificial intelligence methods, which can better read body language and facial characteristics. It is especially useful in colleges or standardized tests, where proctors do not automatically warn large numbers of test users of signs of cheating.

4.8 CONCLUSION

As biometric technology for business and customers worldwide continues to be embraced, businesses will need to become more conscious of the threats and challenges. When there is a biometric data breach, the consumer's confidence in the product is not only severely impaired, but also catastrophic for the businesses concerned. Many solutions help improve biometric protection and mitigate these risks. Figure 4.2 summarizes the main taxonomy of biometric technologies.

Multi-factor authentication – which blends biometrics with other authentication methods such as PIN – is a solution for several businesses. For example, some companies use Iris scanning technology only as part of a multi-factor authentication scheme. Biometrics could replace two-factor authentication as a smoother and faster method and free from some of the challenges of other systems.

FACE RECOGNITION: ROBERT JULIAN-BORCHAK WILLIAMS CASE

In January 2020, while working at an automotive supply company, Robert Julian-Borchak Williams was arrested by a Detroit Police officer in Farmington Hills, Mich. The police wouldn't say why he was being arrested, only showing him a piece of paper with his photo and the words "felony warrant" and "larceny."

The police got his mug shot, fingerprints and DNA, and he was held overnight. Two detectives took him to an interrogation room and placed three pieces of paper on the table, face down. The interrogator asked him if he was in "Shinola," an upscale boutique, where five timepieces worth $3,800 were shoplifted. They showed him an image from a surveillance video, showing a heavyset man, dressed in black and wearing a red St. Louis Cardinals cap, standing in front of a watch display and a second image that was a close-up. The photo was blurry, but it was clearly not Mr. Williams. He picked up the image and held it next to his face.

"No, this is not me," Mr. Williams said. "You think all black men look alike?"

In fact, Mr. Williams was a victim of false positives and errors in facial recognition systems that, according to the research, are not sufficiently accurate for non-white demographics. Facial recognition systems have been used by police forces for more than two decades, yet the case of Mr. Williams is certainly not an isolated incident. This example could also shed light on why Amazon, Microsoft, and IBM have announced that they would stop or pause their facial recognition offerings for law enforcement, according to the New York Times. However, the companies offering face recognition technology to other companies are not necessarily Amazon or IBM for police department [6].

5 Firewall Technologies

5.1 INTRODUCTION

The number of Internet users across the globe marked 4.13 billion in 2019 [1]. Consequently, billions of devices are now connected to the Internet, used for browsing over 1.6 billion websites. With the increasing dominance of the virtual world over the real one in almost all spheres, safety and security is a major concern.

A firewall is one of the most important components of network security. Firewalls serve as a barrier or wall between two networks and can be implemented in software, hardware, or cloud-based applications. Each implementation has its own advantages and disadvantages. Thus, by definition, a firewall filters traffic based on the criteria set by policies that are decided by a network administrator.

Hardware firewalls are often integrated with a router that sits between a computer and a modem. Software firewalls are applications installed on individual computers.

This chapter will begin with a brief historical background, then will review firewall technology, explain how the technology works, advantages and disadvantages of using a firewall, and the different products which use this technology.

5.2 BRIEF HISTORICAL BACKGROUND

The earliest examples of firewalls were implemented on routers, which were the gateways of network traffic. Packet headers (the portion of an Internet protocol (IP) packet that precedes its body and contains addressing) were copied into the Random Access Memory (RAM) of the routers and checked against the device security policy that were based on IP. Internet Control Message Protocol (ICMP) could be used to send and control appropriate error messages. As technology became more advanced, products started using Transport Layer Security (TLS) and Secure Sockets Layer (SSL), which are cryptographic protocols, to provide secure communications over a computer network that was implementing a firewall [2–4].

The first firewall proposal came in 1989 by Jeff Mogul of Digital Equipment Corporation. AT&T labs discovered new technologies like a packet-filtered firewall – routers that operated in the low levels of a network protocol stack and examined each packet that crossed the firewall and tested it according to a set of previously determined rules – and a stateful firewall, which monitored the full state of active network connections.

Firewalls became widespread along with the growth of Transmission Control Protocol (TCP) and the IP.

In 1996 new technologies like Squid and Snort began the commercialization of firewall technologies. As companies emerged, hybrid features were added to firewalls such as Virtual Private Network (VPN), Quality of Service (QoS), antivirus or filters, and Web Application Firewall (WAF) [5].

DOI: 10.1201/9781003038429-5

In 2006, WAF appeared as a stand-alone solution for web applications using HyperText Transfer Protocol (HTTP). With growing enterprise networks, the invention of Internet Protocol version 6 (IPv6) led to the discovery of next-generation firewalls. IPv6 adjusted helps emerging cloud technologies be deployed on various types of cloud deployment features such as public cloud, private cloud, and hybrid cloud. Firewalls were even deployed on hyper-converged infrastructure, which used virtualized network platforms.

TRANSPORT LAYER SECURITY (TLS) AND SECURE SOCKETS LAYER (SSL)

SSL protocol was developed to enable eCommerce transaction security on the web, which required encryption to protect customers' personal data, as well as authentication and integrity guarantees to ensure a safe transaction. Thus, the SSL protocol was implemented at the application layer, directly on top of TCP, enabling protocols above it (HTTP, email, instant messaging, and many others) to operate unchanged, while providing communication security when communicating across the network. When SSL is used, a third-party observer can only infer the connection endpoints, type of encryption, as well as the frequency and an approximate amount of data sent, but cannot read or modify any of the actual data. When the SSL protocol was standardized, it was renamed TLS. Thus, many people still use the TLS and SSL names interchangeably. However, they are different because each uses a different version of the protocol.

HyperText Transfer Protocol (HTTP)	Protocol version 6 (IPv6)
HTTP is a protocol for data communication for the World Wide Web. HTTP allows the fetching of resources.	Devices over the Internet are identified through their IP address. The current IP system is Version 4 (IPv4), which makes available over four billion IP addresses. However, the huge increase in Internet users and devices worldwide means that IPv4 addresses are running out. IPv6, the next-generation protocol, provides approximately 340 undecillion (a number equal to 1 followed by 36 zeros) IP addresses, ensuring the availability of new IP addresses far into the future and promoting the continued expansion and innovation of Internet technology.

HyperText Transfer Protocol Secure (HTTPS)	Virtual Private Network (VPN)
HTTPS is an extension of HTTP. It is used for secure communication over a computer network. In HTTPS, the communication protocol is encrypted using TLS or, formerly, SSL.	VPN is a private network across a public network and enables users to send and receive data across shared or public networks as if their computing devices were directly connected to the private network. Applications running across a VPN may therefore benefit from the functionality, security, and management of the private network. Encryption is common, although not an inherent, part of a VPN connection.

Quality of Service (QoS)

QoS is the description or measurement of the
overall performance of a service, such as a
telephony, computer network, or a cloud
computing service, particularly the
performance seen by the users of a network.
To quantitatively measure the QoS, several
related aspects of the network service are
often considered, such as packet loss, bit
rate, throughput, transmission delay,
availability, jitter, etc.

5.3   HOW FIREWALL TECHNOLOGIES WORK

ANALOGY

To better understand a firewall, consider this analogy. A firewall is like a
strong wall surrounding a city. The wall prevents people and merchandise
from getting in and out of the city after inspection. There might be multiple
inspectors checking people and packages that want to get in or out – but
each inspector checks the same way based on city governance policy. For
example, all people must be checked for proper vaccinations; all packages
are checked to ensure they do not contain drugs. Similarly, when a firewall
is set it controls the system based on the policy that IT professionals set
(Figure 5.1).

The main objective of a firewall is to watch all inbound and outbound traffic and
check if it matches certain predefined rules (firewall policy). If the traffic is in accor-
dance with the firewall policy, it is permitted. Otherwise, it is dropped.

A firewall blocks broadcast addresses, which is the key component used by every
attacker. Operating a system without enabling a firewall is like keeping the front
door of a house open. A firewall prevents any malware trying to install dangerous
software like a trapdoor, which may lead to malicious data collections. A firewall
also helps in the elimination of man in the middle attacks, Denial of Service (DoS)
attacks, and TCP syn flood attacks, where the attacker tries to send too many syn
requests to a single host, which lowers the victim server's performance.

Firewall technologies include Network Address Translation (NAT), packet-filtering,
and VPNs. A firewall controls access to a network by setting up a rule to allow or deny
packets. A firewall offers protection for unauthorized access to confidential data.

Firewalls can identify users according to permissions and validate them to allow
or deny access. A VPN creates a secure private network tunnel for an organization,
which makes the connection more secure. The objective of using a proxy server is to

FIGURE 5.1   Firewall.

act as a cache and make the system faster and because proxy servers are application-specific, they improve aspects of security.

Each request by a network is intercepted by the firewall and checked to confirm it is a valid request. It allows only legitimate or intended traffic from the Internet, while malicious traffic requests and data packets are blocked, thus protecting computer networks from hostile intrusions. Some examples of firewalls are pfSense, ZoneAlarm, Cisco ASA, SonicWall, and the default windows firewall in Windows Operating systems.

5.4 FIREWALL TECHNOLOGIES

The name firewall describes a common architectural design of placing a wall between two entities. This is a useful way to think of what a security firewall does. It acts as a barrier that helps in network traffic flow control both into and out of an organization's network or between different regions of an internal network.

A firewall can serve as the first line of defense in the network and can be utilized for blocking inbound packets of specific types from reaching the protected network. This is known as ingress filtering, and it can be used to reduce the load on high-level firewalls.

The objective of firewalls is to eliminate unauthorized access to data and defend the network. Most of the transactions happen over the Internet. Every organization has its own private network that includes its own active directory domains, Domain Name System (DNS), Dynamic Host Configuration Protocol (DHCP) servers. Attackers use malicious programs that can damage computers and other electronic devices connected to a network. Thus, the enterprise network has to be secured from intruders. Firewalls block any suspicious access from the attackers trying to gather confidential data.

Domain Name System (DNS)	Dynamic Host Configuration Protocol (DHCP) Servers
DNS is a database of internet domain names. The domain name system maps the name people use to locate a website to the IP address that a computer uses to locate a website.	A DHCP server dynamically assigns an IP address and other network configuration parameters to each device on a network so they can communicate with other IP networks.

The three basic types of firewall are (Figure 5.2 shows taxonomy of the firewall technologies):

- Stateless firewalls or packet-filtering
- Stateful inspection or dynamic packet inspection
- Proxy server firewalls or application proxy

5.4.1 STATELESS FIREWALLS OR PACKET-FILTERING

As the name suggests, all the packets (collection of data that can be used by computers) which enter or leave a network are monitored in a packet-filtering firewall. Packet-filtering is a firewall technique used to control network access by monitoring outgoing and incoming packets and allowing them to pass or not, based on the source and destination IP addresses, protocols, and ports.

Each data packet is compared with a predefined set of filter boundaries and regulations and the packet is either allowed, denied, or dropped. This kind of filtering technique works on the TCP and IP layer of the TCP/IP protocol. A packet-filtering firewall does not store the states of the previous data packets and is thus also referred

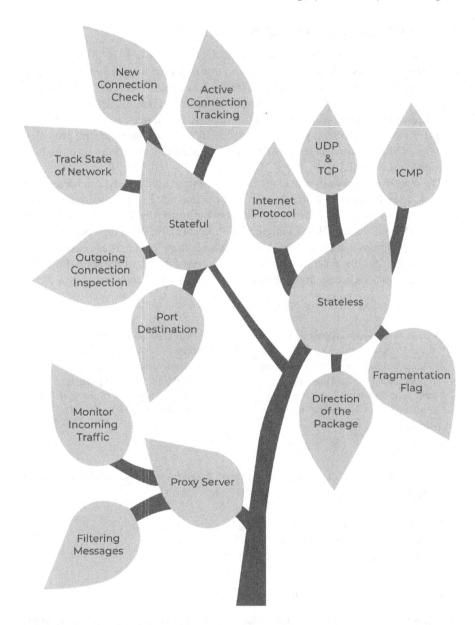

FIGURE 5.2   Taxonomy of firewall technologies.

to as a stateless firewall. Packet filters generally allow or deny network traffic (data packets) based on the following:

- IP addresses of the source and destination
- A protocol like User Datagram Protocol (UDP), Transmission Control Protocol (TCP), or Internet Control Message Protocol (ICMP)

- TCP or UDP port addresses of the source and destination
- Flags present in the TCP header, i.e., ACK, SYNC, CLOSE, etc.
- Fragmentation flag present in the IP
- Direction of the packet

All leading routers used to allow the ability to configure IP datagram filters. Packet-filtering is also supported by Unix operating systems, and the support for using their IP chains is, by default, included in the Linux kernel (key components of the operating system).

5.4.2 STATEFUL FIREWALLS OR DYNAMIC PACKET INSPECTION

Stateful firewalls monitor the connections and filter traffic using a dynamic state table and a firewall ruleset. These firewalls allow traffic that matches a preexisting entry in the dynamic state table for all existing connections and for open new connections. The firewall ruleset is referred to for deciphering whether to accept or deny entry. This enables the internal hosts to initiate connections to external hosts by means of an arbitrary port number. On the other hand, those external connections which have not been initiated by internal hosts, or the ones which do not match the allowed rules in the ruleset, are filtered out.

Stateful firewalls are an advanced and modern extension of stateless packet-filtering firewalls. They can continuously keep track of the active connections they have, such as Transmission Control Protocol (TCP) streams or User Datagram Protocol (UDP). Additionally, they also keep track of the state of the network. They also have the upper hand over their stateless cousins by having an ability to acknowledge and utilize the incoming traffic and data packets, which allows them to differentiate between trusted/reliable and malicious traffic or packets. Due to this advantage, stateful firewalls possess one of the most powerful security tools that protect their network connections: that a new connection will need to introduce itself to the firewall, before being permitted into the list of established connections.

For example, a stateful firewall could be configured in such a way that it would allow only outgoing connections to destination ports 80 (HTTP) and 443 (HTTPS). These are the standard ports used for servers all over the Internet. Thus, this firewall setting could be considered to only give inside hosts located behind the firewall the ability to start a connection to the external world.

5.4.3 APPLICATION PROXY/PROXY SERVER FIREWALLS

Application proxy firewalls are considered to be the most secure type of firewall because they prevent direct network contact with other systems. Application proxy firewalls work at the application layer of the TCP/IP convention stack, giving proxy services to a set of predefined applications. Every application proxy sits between the internal network and the outside world. Instead of creating a direct communication link with the outside world, packets travel between the external world and the proxy. Application proxy firewalls give a high level of security and logging capabilities.

An application proxy resides in between the secure network and the network from which additional security is required. The application intercepts each request to the destination; it then starts its own request and the destination server replies to this application proxy request by considering it to be the destination host. In this way, the source and the destination host never communicate directly, making it much more secure. This indirect communication link generation also allows the addition of security checks and protocols; however, that will be at the cost of the throughput. Furthermore, the methodology allows the verification of a complete data packet, including its application part, and thus, can be considered as one of the most secure types of firewalls. Lastly, it provides additional functions like content caching.

5.4.4 CLOUD FIREWALLS

The rapid expansion of cloud computing businesses means that many firewalls are also shifting into the cloud. Cloud firewalls manage the flow of information between outside domains and internal systems and protect the operation's data. Cloud firewalls work like an on-premise firewall except that they're in the cloud, hosted by a service provider.

5.5 ADVANTAGES AND DISADVANTAGES OF FIREWALL TECHNOLOGIES

Firewalls are a necessity rather than a luxury now. They are attached as external hardware or software. Their necessity depends on the exact needs of a network and the extent to which the firewall is required. Levels can be decided based on the user's personal level to an external network level. The place where a firewall is attached becomes where each and everything is monitored and filtered accordingly.

5.5.1 ADVANTAGES

Firewalls are advantageous in the following ways, based on how they are implemented:

- Firewalls check for each and every packet, which is either inbound to the network or outbound from the network. Rules make the network secure by forwarding authorized packets and blocking unauthorized packets. This filtering of packets makes firewalls secure.
- Spoofed source addresses can be easily tracked down by packet-filtering firewalls.
- Because they only check current packets, firewalls can perform quickly.
- Dynamic inspection firewalls keep a log of all the packets that have passed through. Therefore, they can track down previous transitions of the packet and find a pattern that can be malicious.
- Firewalls make sure to remove all connections which are not well established and improperly configured.
- The restriction of websites in a certain network is possible, which provides better security both for inbound and outbound traffic.
- New firewalls are now also capable of load balancing, thus making network distribution better and preventing congestion.

5.5.2 Disadvantages

The issues that can cause issues to administrators or even users when using firewalls include:

- Installing and maintaining firewalls has a financial cost leading to an increase in capital and operational expenditures.
- Every packet needs to be passed from a network firewall, thus forming a single point of failure.
- An unplanned network infrastructure can cause more complex firewalls to be built, thus causing congestion.
- Firewalls may block trusted services like VPN. Therefore, it may be necessary to reconfigure firewalls.

5.6 CONCLUSION

The use of the Internet has increased exponentially in the last few years, with almost 4.13 billion people a part of this ecosystem now [1]. With an increase in online users, privacy is at stake, and security becomes much more important. Firewalls are considered to be one of the security measures that can help in preventing identity theft and privacy disclosure.

Firewalls give us the security we need by filtering the authenticated data from an unauthenticated one. The three main types of firewalls discussed in this chapter were packet-filtering (stateless), stateful filtering, and application proxy firewalls. This chapter reviewed how firewall technologies work and how they protect internal networks.

Firewalls are crucial for the protection of an internal network from specific kinds of malicious data packets. Advanced firewalls also provide the capability to protect a network from unauthorized remote access and by creating a security layer between the secure network and the outside world. Figure 5.2 shows the taxonomy of firewall technologies.

DISRUPTED FIREWALLS AT U.S. POWER UTILITY

In March 2019, a denial-of-service (DoS) attack that exploited a known vulnerability in a firewall caused "interruptions of electrical system operations" and caused disruptions at an unnamed utility in the western part of the United States. The incident impacted California, Utah, and Wyoming, but it did not result in any power outages.

According to E&E News report, the North American Electric Reliability Corporation (NERC) revealed that the incident involved a vulnerability in the web interface of firewalls used by the impacted organization.

The failure seemed to be related to firewall security updates and the lack of a proper firmware review process to vet security updates before being deployed [6–8].

6 Virus Detection

6.1 INTRODUCTION

After the introduction of computers and distributed systems, the first-ever virus attack took place in the mid-1980s. Since then, thousands of new computer viruses have appeared. According to CNN, nearly 1 million new malware threats are released every day [1].

Antivirus software for computers, also known as anti-malware software, is used to prevent, detect, and remove malicious software. This software helps a user quarantine the infected file. Antivirus software has evolved over the years. It provides users with protection from modern digital threats like ransomware, keyloggers, rootkits, trojan horses, phishing attacks, and botnet DDoS attacks. Malware itself is also evolving every day. In recent years, the proliferation of malware has posed a serious threat to computer systems worldwide. New, sophisticated and more complex viruses are posing major problems to traditional static virus detection techniques. Viruses have better and ever-increasing opportunities to spread with the increase in interconnectivity and interoperability in computer systems.

After a brief historical background, this chapter will review virus detection technology, explain how the technology works, advantages and disadvantages of the different virus detection technologies, and the different products that use the technologies.

6.2 BRIEF HISTORICAL BACKGROUND

In early 1971, the first-known computer virus surfaced and was known as the "Creeper virus." It infected the DEC's PDF-10 computers (a mainframe computer family), which had TENEX OS running on it. Consequently, a program to delete this virus was developed by Ray Tomlinson called "The Reaper," which is considered the first virus detection and elimination software ever written. The Reaper was a virus itself and was actually designed to spread and find the creeper, and delete it.

In 1983, Fred Cohen used the term "computer virus" and described it as a program that infects other computer programs and attempts to modify them, after which it copies itself [2, 3].

In 1986 the first widespread computer virus appeared, called "In the wild," and this was followed by an exponential increase in viruses. In days before the Internet, viruses spread mostly through floppy disks. During this time, virus detection and elimination programs and systems started becoming developed and were famously known as Antivirus software.

During these times, antivirus software mainly evaluates ".exe" (Executable Files) files and boot sectors of inserted floppy disks and hard drives. With the growth of the Internet, viruses began to spread online.

DOI: 10.1201/9781003038429-6

G Data software introduced the first commercial antivirus product for the Atari ST home computers. The Ultimate Virus Killer soon became the most common industry standard for Atari systems. Around the same time, John McAfee founded the McAfee Antivirus Software company.

In 1991, the Computer Antivirus Research Organization introduced the "Virus Naming Scheme," which still remains today as the standard for cybersecurity companies. In 1991, Symantec released the Norton Antivirus. AVG Technologies released the Antivirus Guard in 1992. F-Secure from Finland is considered to be among the first antivirus programs that became famous on the Internet.

In 1991, the European Institute of Computer Antivirus Research (EICAR) was founded to research and develop virus detection and extermination techniques and methods. The first open-source virus detection and elimination project was founded in 2000 and called the OpenAntivirus Project. ClamAV was the first open-source anti-virus application.

As more viruses got introduced, they became more sophisticated and difficult to detect. It became necessary for virus detection systems to employ different strategies and detection algorithms. They also had to accommodate checking an increased variety of files.

The first cloud-based antivirus was proposed after noticing that most users were consistently connected to the Internet.

The traditional signature-based virus detection techniques are considered ineffective today. Thus, the antivirus industry has seen a paradigm shift and a move to adopt signature-less methods for detecting viruses.

6.3   HOW VIRUS DETECTION TECHNOLOGIES WORK

ANALOGY

Virus detection can be understood by thinking about the flu or flu-like viruses. If a virus is going around and making people sick, a city might protect its people by testing people for all types of known viruses at the city's entry points. If a vaccine exists for the type of infection going around, people may be required to vaccinate before entering the city. Otherwise, they may be required to quarantine until the risk of viral infection has passed. Virus detection applications automatically do this for computer viruses (Figure 6.1).

6.3.1   Virus Scanning and Detection Process

All program files newly entering a system are scanned by the virus detection system, generally known as the antivirus. The ones that match the virus signatures are flagged as viruses or infected files or programs and are blacklisted or quarantined. The ones who have passed this round are tested with the Host Intrusion Prevention System, wherein known files or programs are allowed to enter the main system and execute whereas unknown ones are further tested in a sandbox (a security mechanism

FIGURE 6.1 Virus detection.

for separating running programs) where they are run in a restricted environment, and their behavior is analyzed for suspicious activity. The ones that have been finally identified as good files are added to a whitelist (people or things considered to be acceptable or trustworthy).

Virus detection technologies provide various features depending upon the type of Operating System (i.e., Windows, macOS, Linux) and the platform (i.e., mobile devices, desktop, or server) on which they operate.

Below is a summary of virus scan types:

- On-demand Scan: Manually scanning devices by the user
- On-access Scan: Automatic scanning of devices by the product itself after creation or modification
- Boot-time Scan: The scans performed at boot time
- CloudAV: Automatic scanning of files in the cloud
- Email Security: Protection of emails from viruses
- AntiSpam: Protection of files from spam, scam, and phishing attacks
- Web Protection: Protection from online malicious programs and identity theft
- Heuristics: The detection of new and modified viruses

- Firewall: Ensures authorized access to and from a private network
- Sandbox: The process of running applications in an isolated environment

The type of antivirus software will define whether it will provide all or some of the above features. By far, the Behavior-Based Malware Detection Technique (BMDT) is considered to provide the most features. BMDT evaluates an object based on its intended actions before it can execute that behavior. Then the object's behavior, or potential behavior, is analyzed for suspicious activities.

When an antivirus software scans a system for any new viruses, it compares the files to known malware. It uses three types of detection methods:

- Specific Detection: The antivirus software looks for the viruses known to it by using a specific set of characteristics.
- Heuristic Detection: The software searches for odd behaviors of the systems or odd file structures. The virus is detected by finding the strange behaviors of a system.
- Generic Detection: Certain variants are assigned to virus families. In this type of detection, the antivirus software looks for the variants assigned to these families.

With the prolific advancement in viruses, traditional and static scanning technologies are facing tremendous limitations in scanning for polymorphic viruses (file infectors that can create modified versions of themselves to avoid detection). Static scanning involves static analysis of a virus and then finding the signature using the strings in the virus. Whenever the antivirus is scanning the new file, it first searches for the virus signature in its virus signature database.

Recently, machine learning methods have been used to detect unknown viruses. Some companies are working on the development of detection methods based on neural networks. This method requires less prior knowledge and less training time as compared to other methods

6.4   VIRUS DETECTION TECHNOLOGIES

The best virus detection systems use a multi-dimensional, multi-layer approach, which achieves a good detection rate. Figure 8.2 shows the main virus detection technology taxonomy.

There are a number of techniques for virus detection that are more often used together than in isolation [4].

6.4.1   SANDBOX DETECTION

This involves analyzing the behavioral fingerprint of programs that could be infected at run time. It runs these programs in a virtual environment in isolation from other applications so that the platform, system, or applications are not affected, and it further logs their actions. Depending on the logs, it determines if the program is infected or not. This is a useful technique but is very resource-consuming and hence rarely used.

6.4.2 DATA MINING TECHNIQUES

Data mining and artificial intelligence techniques use machine learning algorithms to classify a file as infected or not. When the antivirus system is trained, it learns to identify the infected ones.

6.4.3 SIGNATURE-BASED DETECTION

Signature-based detection is the most common virus detection technique, and it uses signatures to identify viruses. In this method, a file identified as a virus is analyzed, and once confirmed, its signature is generated from the file. A database of signatures is maintained, and during scanning for viruses in a system, files are looked up to match with a signature of that of a software.

Signature-based detection is becoming increasingly ineffective, as viruses today are metamorphic, which encrypt different parts of themselves in order to disguise and not match virus signatures.

6.4.4 HEURISTIC DETECTION

In heuristic detection, viruses are classified into a family. Generally, viruses are derived from a previous virus after modifying them. Hence most viruses match mainly with a generic signature. It is quicker to detect the virus family than the exact virus, which is sufficient enough to detect and eliminate it.

6.4.5 REAL-TIME PROTECTION

Real-time protection involves background guard, auto-protect, on-access scanning, resident shield, and other techniques that provide protection in real time. Real-time protection constantly monitors systems and aims to protect from viruses and other malware. It monitors for suspicious activities such as connections to the Internet, connecting with external devices such as USB, CD, or Bluetooth. Real-time protection also monitors newly downloaded programs.

6.5 ADVANTAGES AND DISADVANTAGES OF VIRUS DETECTION TECHNOLOGIES

6.5.1 ADVANTAGES

In general, antivirus technologies have the following benefits:

- Safeguard browsing activities to identify and eliminate threats.
- Provide protection from cybercrimes such as phishing attacks.
- Enabling parental control to ensure appropriate activities take place within the home or organization.
- Scanning of removable devices to obstruct incoming viruses.
- Related technologies can prevent more specific and complex problems, such as heuristics scanning.

- Sandboxing Technique can give a complete description of behaviors and intents of malware.
- Real-time protection helps in recognizing Zero Hour threats.
- AVG antivirus (Antivirus Guard) can detect replicating viruses at boot time.

6.5.2 DISADVANTAGES

The challenges and disadvantages of virus detection technologies are:

- Costs incurred due to unforeseen renewal plans mentioned in the end-user license agreement of commercial software
- Failure of non-signature-based methods in identifying new viruses
- Decrease in effectiveness of technologies with time, as virus designers are well aware of the workings of antivirus applications and can gain an edge in tricking antivirus companies
- The masquerading of malware applications as legitimate antivirus software (e.g., Mac Defender and Microsoft Antivirus).
- The false recognition of files as malware resulting in their deletion or destruction and the ensuing costs
- Detection of rootkits (a collection of computer software) due to their permissions and privileges that allow them to neutralize the antivirus software without being visible.
- Restoring damaged files can sometimes only be done through backups or shadow copies.

6.6   APPROACHES USED IN VIRUS DETECTION

Data mining techniques play an important role in virus detection. Smartphones, mobile devices, and various handsets are used today. These technologies, along with behavioral-based analysis, help in detecting a virus.

Signature-based detection is most commonly adopted by the corporate sector. Hardware appliances are adopting this technology frequently in offices rather than in personal households. Integration of this signature-based detection, firewall, and specialized IT skills ensure a well-secured system for any corporation.

Industries which require work to be done as fast as possible and want to remove the window of vulnerability most often use heuristic detection. In addition, audio-video companies prefer heuristic technology since it helps their systems to detect what a piece of code is going to do in the future and decides an appropriate action, whether it is desirable or not.

Windows Defender and many other personal antispyware programs make great use of real-time protection. Therefore, this technology ensures that the system can be protected at any point of time during which it is used. As such, it is commonly used in the banking sector where the protection of customer data is crucial at each and every step of transactions.

6.7   CONCLUSION

Antivirus technologies ensure that users can protect themselves and their devices against malware or cyber-attacks. They act as a warning system against many types of attacks and can keep hackers away. They play a vital role in securing information and data, therefore protecting them from getting compromised. Various technologies

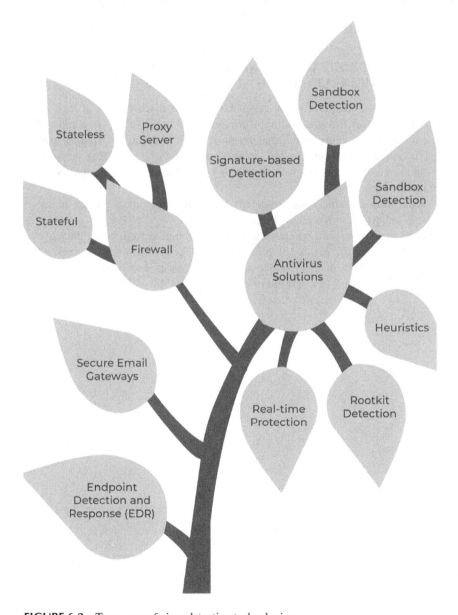

FIGURE 6.2   Taxonomy of virus detection technologies.

can be used to detect virus attacks of a system such as Sandbox detection, data mining techniques, signature-based detection, heuristic detection, and real-time protection detection. There is great functionality associated with each technology. Each of the technologies comes with some pros and cons, but all in all, they prove to be a great asset. Due to their high functionality, these products are becoming increasingly adopted day by day, ensuring complete security of systems and protecting users from hackers. Figure 6.2 shows the taxonomy of virus detection technologies.

STUXNET: A WEAPONIZED COMPUTER WORM ORIGINALLY AIMED AT NUCLEAR FACILITIES

In June 2010, a computer worm called Stuxnet was discovered in at least 14 industrial sites in Iran, including a uranium-enrichment plant. A computer worm has the capability of spreading on its own without the need of user interaction, and Stuxnet was able to accomplish this by spreading between Windows computers through workers using USB thumb drives connected to an infected machine.

This is an extremely sophisticated and thought-out attack which many experts believe was created by the United States.

In its earliest version, Stuxnet was used to sabotage centrifuges at the plant for about a year, increasing the pressure inside them to damage the devices. These attacks continued to infect several companies and plants and were deemed to be a mark of a new era of cyber warfare [5].

7 Phishing Detection

7.1 INTRODUCTION

Phishing originated in 1996 and referred to a cyber fraud crime in which an attacker achieves certain malicious purposes, such as the intrusion of an internal network, gaining sensitive information and data, or malware deployment. The act of pretending to be a legitimate source by deception is termed as phishing, which is considered to be an attack based on social engineering.

Phishing attacks are evolving every day, with everyone from individuals to large businesses being targeted and the result of billions of dollars of financial losses every year. Phishing attacks account for more than 80% of reported security incidents [1]. In 2015, an estimated $4.6 billion in financial losses were attributed to phishing attacks. The number of phishing attacks is also increasing every year [2] through junk email, instant messaging tools, mobile phones, short messages, or web pages that send false advertising and other deceptive information from banks and other well-known institutions. The intention is to induce the user to log into what looks like a fake website that looks very real to them and have them input sensitive information, such as the user's name, password, account ID, Social Security Number, ATM PIN, credit card, etc.

Research has shown that even an experienced user can fail to distinguish between legitimate and illegitimate websites due to high resemblance. Attackers have become increasingly skilled at replicating websites not only by mimicking the color and format, but also the text of the websites they're impersonating. They do this by copying HTML source code from the original websites [2].

While the phishing model is evolving, in order to identify and prevent attacks, phishing detection technologies use some of the very same characteristics that attackers use.

After a brief historical background, this chapter will review the three main types of phishing detection technologies based on visual similarity, blacklist, and web crawling. We will explain the main phishing detection technologies and how they work, the advantages and disadvantages of each, and the different products which use phishing detection technologies.

7.2 BRIEF HISTORICAL BACKGROUND

In 1990, one of the first victims of a phishing attack was American Online Networks (AOL) [3]. AOL did not validate the fraudulent users and their fake credit card numbers, and these users got away without paying their ALOL bills. After the company overcame this problem, attackers developed new ways to get AOL accounts fraudulently. Attackers sent emails and messages to legitimate users acting as AOL employees and asked them to confirm their usernames and passwords.

The expansion of online activities and particularly eCommerce banking brought with it an increase in phishing. The first known phishing attack on eCommerce

DOI: 10.1201/9781003038429-7

websites started with the E-Gold website in June 2002 [4]. Hackers then started registering several new domains that resembled names of popular sites like eBay and PayPal and sent their customers spoof emails that tricked them into providing their credit card details and other personal information.

By early 2004, phishing evolved into a profitable business, and hackers started attacking banks, enterprises, and their customers. Phishing techniques also evolved and became more sophisticated by using various ways to access victims' data including:

- Spear phishing: sending emails ostensibly from a known or trusted sender to induce targeted individuals to reveal confidential information
- Vishing (voice phishing): using a telephone to scam the user into surrendering private information that will be used for identity theft
- Smishing: tricking people into giving up their private information via a text or SMS message
- Keylogging: monitoring software designed to record keystrokes made by a user
- Content injection: inserting malicious content into a legitimate website
- Content spoofing: manipulating what a user sees on a site by adding parameters to its URL

To combat ever-increasing email spamming in 1997, Dave Rand and Paul Vixie (a computer scientist who contributed to the Domain Name System (DNS) protocols in the early stages of the Internet), created Domain Name System-based Blackhole Listing (DNSBL) also known as DNS-BlackListing.

With time, phishing detection technologies based on web crawlers appeared. A web crawler is an automated script that browses the World Wide Web in a methodical, automated manner. This process is called web crawling or spidering. Many legitimate sites, in particular search engines, use spidering as a means of providing up-to-date data.

NASA developed the first web crawler in a program called Repository Based Software Engineering (RBSE). It was developed in 1994 at the University of Houston to collect indexes and run statistics on the web.

7.3   HOW PHISHING DETECTION TECHNOLOGIES WORK

ANALOGY

Phishing detection is like a police investigation of a suspect. Once the individual is considered a suspect, the police officer might look at the individual's personal history, driver license record, criminal record, and maybe even international record to see if there are any red flags about that person. Phishing detection technologies work similarly. Anytime the technology identifies a suspicious connection, it will block the potential phishing attack. However, this will not work if there are no red flags on record for that individual or connection in the case of a phishing URL (Figure 7.1).

FIGURE 7.1 Phishing detection is like a police investigation of a suspect.

Phishing always has a strong relationship with its target, and there is a certain amount of misleading information. For example, phishing domain names are similar to legitimate links and visually similar content to induce users to enter sensitive information. Phishing detection analyzes this content (URL, mail, web page, etc.), which is misleading information, to detect and identify phishing.

One of the most important features of a phishing attack is the visual similarity. Attackers try to replicate legitimate domain names, exact websites, and email formats to trick victims. Researchers have been studying phishing attacks and devising various techniques to identify such attacks. The similarity is also utilized to differentiate between legitimate and illegitimate emails and websites. The visual representation of legitimate websites is stored in a database. Whenever a third-party website crosses the similarity threshold, a website is identified as a phishing site [2].

7.4 PHISHING DETECTION TECHNOLOGIES

In recent years phishing has become the most effective method for cyber-attacks and its technology has evolved. This section will discuss the main phishing detection technologies.

Figure 7.2 shows the taxonomy of the main phishing detection technologies.

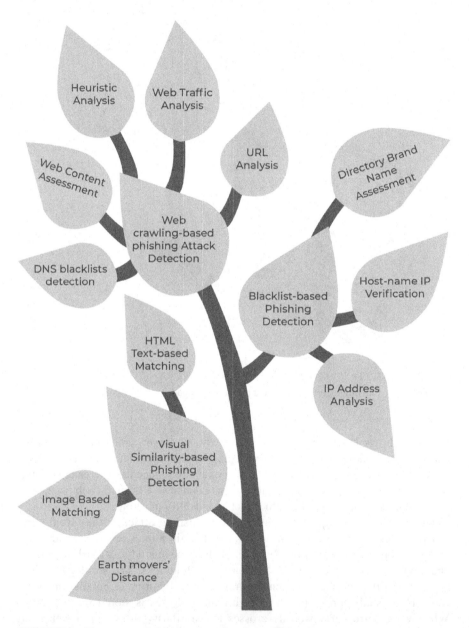

FIGURE 7.2 Taxonomy of phishing detection technologies.

7.4.1 VISUAL SIMILARITY-BASED PHISHING DETECTION

Visual similarity-based phishing detection can be HTML text-based matching or image-based matching [2]. The Document Object Model (DOM) treats a HTML document as a tree structure that divides web pages into blocks according to visual cues. It then uses three metrics to evaluate the visual similarities between the website to be detected and the legitimate website: block-level similarity, layout similarity, and style similarity.

Visual similarity-based phishing detection largely depends on the results of web page segmentation, especially the calculation of block-level similarity and layout similarity. Therefore, the detection effect of this method relies on the availability of DOM representations and cannot detect similar appearances, but the DOM represents a different web page [6].

Earthmovers' Distance (EMD) is a method to measure the visual similarity of web pages. This method maps web pages to low-resolution images for the first time, then uses color features, and coordinates features to represent the features of the image. EMD is used to calculate the feature distance between web page images and an EMD.

The EMD method is completely based on the web page's image features and does not depend on the availability of HTML content. However, due to the significant number of suspicious and legitimate web pages, irrelevant web page image pairs may also have extremely high similarity, leading to an increase in false detection rates [2].

Visual similarity-based phishing detection technologies target the visual resemblance of the phishing sites. The visual representations of the legitimate websites are stored inside the database, and whenever a site in question crosses the similarity threshold, it is designated as a phishing site. If there is an extreme similarity between the legitimate site and the site in question, the site in question is likely pretending to be a legitimate website and trick users.

Although this technique is quite effective in identifying deceptive phishing attempts, there are always false positives and false negatives. Hence this technique cannot guarantee 100% accurate identification of phishing sites. Some phishing websites may pass through, and some legitimate websites may be flagged. Also, if attackers replicate just enough of the visual appearance of a legitimate website and manage to stay under the similarity threshold, they might be able to get away with it.

7.4.2 BLACKLIST-BASED PHISHING DETECTION

Blacklist-based phishing detection focuses on suspicious Uniform Resource Locators (URL) and matching those to legitimate URLs. To avoid detection of a blacklist, phishing attackers often change the URL of their phishing page, and any such change can lead to the failure to match the URL to the blacklist, resulting in a missing detection.

To address the limitations of exact matching, Prakash et al. have proposed an improved method called PhishNet [5]. This uses two components. In the first, the authors propose five heuristics (such as universal top-level domain name substitutability, directory structure similarity, etc.) to enumerate simple combinations of known phishing sites to discover new phishing URLs. The second component consists of an approximate matching algorithm that dissects a URL into four parts: IP address, hostname, directory structure, and brand name; and attempts to approximately match those with the corresponding part of the blacklist in order to determine if it's a phishing URL. Thus, PhishNet can expand the blacklist and detect phishing that had not appeared on the blacklist.

Felegyhazi et al. [6] explore proactive domain blacklisting. This method is based on registration and name server information and leverages the key observation that Internet cyber criminals need to register a large number of domain names to maintain their activities. A blacklist of domain names is used as a sub-list, and the list is augmented with name system information in the DNS zone file and the WHOIS database, which contains domain name registration information. At the same time, the method also makes use of name server registration characteristics, such as freshness and self-analysis.

Most blacklist phishing detection technologies maintain a database of approved and unapproved URLs. The database can also contain the IP address of a known phisher. If a sender's IP or ISP has been blacklisted, users will be unable to send emails from that address using various email servers (i.e., Gmail, Hotmail, etc.). Moreover, the blacklisted person will not be able to receive emails on the blacklisted IP.

A URL or an IP address can be added to the blacklist in three ways. First, a person can report an IP, a URL, or an email address as a phishing attack. Second, improper email behavior can also result in being added to the blacklist. Each ISP server has a rate per hour of emails that can be sent from an IP address. If such a limit is exceeded, then there's a risk of being added to the blacklist. Third, the number of recipients per message, per hour, and the number of connections also play a role in being added to the blacklist. Additionally, the number of non-deliverable messages can contribute to being blacklisted. Finally, DNSBL (mentioned in the history section) can also be a factor in being placed on the blacklist.

Blacklist technologies do not have an automated way to whitelist IPs. Legitimate email senders using a new IP address that does not yet have a reputation can be blacklisted. Furthermore, blacklisting technologies might not automatically add new URLs that were used for phishing attacks.

Blacklist technologies are very labor-intensive. They do not implement automatic and machine learning-based methods for detecting new suspicious IP addresses, email addresses, and URLs for phishing attacks.

7.4.3 WEB CRAWLING-BASED PHISHING ATTACK DETECTION

A Web Crawling Phishing Attack Detector (WC-PAD) can check the websites visited by a user. When the user visits a website, the first thing that WC-PAD checks is whether the website link is present in the DNS blacklists website list or not. If any link in that list matches the website link, then a warning will be sent to the user that the website is not safe [7].

If the website link does not match any link, then the software will crawl through the web page and check every link present on that website to find faults in the web index. If any links on the web page are found to be wrong, then a warning will be sent to the user. A web crawler is also used to get important information from a web page such as the web content features, URL features, and web traffic features.

Another technique of phishing detection is heuristic analysis. WC-PAD checks three things: web content, URL, and web traffic.

- In web content, the web crawler goes through all the links and copyrights present on the web page. Based on the information found by the web crawler, the web content is analyzed to see if the content is legal or illegal. If any content is found to be illegal, then a warning will be sent to the user.
- In a URL analyzer, WC-PAD extracts the website URL details and checks if the URL is correct and legal or not. The URL is mainly divided into these parts: <protocol>://<SubDomain>.<PrimaryDomain>.<TLD>/<PathDomain>. In the example of: http://paypal.abc.net/index.htm, in this URL, the protocol is HTTP, SubDomain is PayPal, PrimaryDomain is ABC, top-level domain TLD is net, domain is abc.net, and PathDomain is index.htm. A URL analyzer checks for the number of times "-" is in a URL. A legitimate website does not have "-" or ";" whereas phishing sites will use these characters frequently. A URL analyzer also checks for the correct dictionary spelling of words and ascertains if there are any misspelled words, with the help of Levenshtein distance (LD), which is a tool that calculates if a word is spelled correctly and if not, the degree of the difference in the spellings. If the distance given by the Levenshtein is much less than 2 or 3 then there might be a possibility of a phishing attack. A web crawler not only extracts the URL of a website, but also all the interconnected URLs from that website, and a URL analyzer assesses if all of those URLs are valid or not.
- In a web traffic analyzer, certain parameters of websites are checked, such as total visits on the website, total per page visit, average visit duration, bounce rate, Google PageRank, and AlexaReputation. Google PageRank and AlexaReputation are calculated based on the number of links from different websites to this website. For example, on a zero-day phishing website, the total visits of the web page and, the average visit duration would be low, since these websites do not last many days. After that, hackers move to a new hosting server. This allows them to evade detection and maintain an ongoing campaign without being detected and blocked. A zero-day attack is when a completely new phishing URL, email address, IP address, and email context are used in the attack.

There are three main things that are done by this technology to detect if the URL is used for phishing or not. The first thing will maintain a list of all the blacklisted websites on the DNS and regularly update that list if new phishing URLs are found. Secondly, to detect a zero-day phishing website that will easily bypass undetected from the first step of this technology. This technology uses a web crawler that crawls the main web page and all the interlinked web pages to collect total visits on the website, total per page visit, average visit duration, bounce rate, Google PageRank, and AlexaReputation. After collecting these things, it will calculate that the page is illegal or not. The third step is the heuristic analyzer that analyzes three main features of the website that are web content, URL, and web traffic. These features play a very important role in distinguishing between an illegal website and a legal website.

7.5 ADVANTAGES AND DISADVANTAGES OF PHISHING DETECTION TECHNOLOGIES

7.5.1 ADVANTAGES

7.5.1.1 Visual Similarity-Based Phishing Detection

One of the reasons phishing is successful is because a phishing site tricks the human brain by copying and resembling legitimate websites. Visual similarity-based phishing detection techniques solve this problem by matching the features of phishing sites against those of the legitimate websites they are trying to replicate.

Visual similarity-based phishing detection techniques help Internet users by doing the tedious task of comparing the websites in question with all legitimate websites. This method rigorously examines and compares underlying HTML DOM structures and layouts of a website in question and provides thorough results.

7.5.1.2 Blacklist-Based Phishing Detection

The benefits of having a blacklist for phishing detection are the guarantee that a known phishing-based URL, IP address, or email sender will be blocked from a mail server. Moreover, the quantity and quality of phishing URLs in a blacklist play a key role in phishing detection. The higher the quality and the more of those URLs in a blacklist, the more effective the phishing detection mechanism will be.

Blacklist-based phishing detection is one of the oldest and most widely used technologies. A blacklist system is easily testable, and the lists can be shared and tested across different platforms and industry partners. Ease of testing will provide a highly robust product. Moreover, since blacklist detection is widely used, installing it in a new system and configuring it is easier compared to more complicated systems.

7.5.1.3 Web Crawling-Based Phishing Attack Detection

The main advantage of using web crawling-based phishing attack detection technology over other technologies like blacklist phishing detection or visual similarity is that it has three-phase detection. A phishing website can bypass one way of checking by making appropriate changes, but in WC-PAD, it is too difficult to beat the WC-PAD software [8].

A phishing website will first have to go through a blacklist test, in which most of the old phishing will be detected, and a warning to the user will be shared, but if the website is new, it can bypass blacklist phishing detection and go undetected. In WC-PAD, that new website will have to phase through a web crawler and heuristic analyzer. These two make it almost impossible for a zero-day phishing website to evade detection.

7.5.2 DISADVANTAGES

7.5.2.1 Visual Similarity-Based Phishing Detection

One of the drawbacks of visual similarity-based phishing detection technology is that it is very time-consuming. It aims to find the underlying block level and layout level similarities. It also takes time to compare the website in question with all websites in a database and to calculate similarity scores for each one. Hence it is also resource-intensive. Also, as this technology highly depends upon the availability of the DOM structure of a web page, it provides a false positive output when the DOM structure is different but the website appearance is the same.

7.5.2.2   Blacklist-Based Phishing Detection

One of the most significant disadvantages of blacklist phishing detection is the inability to detect zero-day attacks. Because such information does not yet exist in a blacklist database, such attacks can pass through filtration systems and reach users' inboxes.

Another disadvantage is the slow manual entry of new phishing data into blacklist databases. Users need to report phishing emails and URLs manually, which means it will take time for databases to be updated and to take effect in filtering future attacks. Moreover, blacklisting is known to have a rate of false positives, since a legitimate website can be flagged and placed into the blacklist, which would then mean that legitimate emails could be blocked from user's inboxes.

7.5.2.3   Web Crawling-Based Phishing Attack Detection

The major disadvantage of using this technology is that it uses many resources and is time-consuming as compared to its counterparts [8].

7.6   WHAT PRODUCTS USE PHISHING DETECTION

7.6.1   Blacklist-Based Phishing Detection

Google's Safe Browsing feature is an example of blacklist-based phishing detection. It is famous for implementing a blacklist of its own that warns users if they are attempting to navigate to a website known for hosting malware. Moreover, it has put in place a practice of warning users of websites that try to install "Unwanted Software" in their machines. Unwanted software is not necessarily malicious; however, it is disguised as a useful software which, in reality, changes the user's browser preferences, such as their homepage and their preferred in-browser search engines. Also, Google Safe Browsing has a list of websites that show signs of social engineering. If a website has a practice of tricking users into revealing their personal information, then such websites are flagged and put under Google's Safe Browsing blacklist. For future visits, such sites will be blocked from a user. Users will see a warning message about the dangers that the site hosts.

Google's Safe Browsing is accessible as an API that developers can use to integrate into their applications and browsers. So not only Google Chrome, but also other browsers such as Firefox, Safari, Vivaldi, and GNOME browsers can make use of Google's Safe Browsing blacklist. This feature is also capable of warning webmasters of attacks that are happening on their own sites. Moreover, Safe Browsing helps developers diagnose a breach and resolve the issues from the attack.

Another example is StopBadware, which is a not-for-profit open-source project that aggregates blacklisted URLs and helps legitimate websites to be removed from blacklists. To date, there are over 1 million URLs that have been blacklisted in the StopBadware database, and there are over 200,000 sites that have been de-blacklisted. The advantage is that anyone on the web, i.e., regular users, can have the capacity to report phishing URLs. Webmasters then have access to the blacklisted database and can use it in their filtering systems. Since it's a community-run organization, there's a high volume of collaboration in terms of building and fine-tuning the blacklist, which results in very low false positives.

7.7 CONCLUSION

This chapter reviewed the leading phishing detection technologies: visual similarity detection, blacklist detection, and web crawler detection. Visual similarity detects phishing in a website based on the visual similarity of the illegal website with the legal website, which is, in most cases, less than 50%.

Blacklist detection takes a different approach to detecting phishing websites by maintaining a list of blacklisted websites on the DNS. It checks a website as soon as a user tries to access it and gives the user a warning if there is a match.

Web crawler phishing detection is a more advanced version of blacklist detection. It has a basic blacklist detection functionality and two additional phases: a web crawler and heuristic analysis that can detect zero-day phishing websites, which can easily go undetected in blacklist only detection.

Phishing methods evolve very fast, and there is a constant need for innovation to detect zero-day phishing websites and make the user experience risk free.

SONY HACKERS TARGETED EMPLOYEES
WITH FAKE APPLE ID EMAILS

In the fall of 2014 Sony Pictures Entertainment was infiltrated by hackers who stole gigabytes worth of files, ranging from emails and financial reports to digital copies of released films. Then just before Thanksgiving, the attackers crippled Sony's PCs with malware that erased the machines' hard drives.

The hackers accessed Sony's networks through a series of phishing emails, which targeted system engineers, network administrators, and others who were asked to verify their Apple IDs (which is Apple's system for identifying users to enable access to Apple services such as the App Store and iCloud). As such, these Sony employees were tricked and convinced to click and verify their accounts. The hackers were successful because the Apple ID verification phishing mails were very convincing [8].

PHISHING USED EMAILS TO STEAL OVER $100 MILLION FROM GOOGLE AND FACEBOOK

In 2016, two conspirators created convincing forgery emails using fake email accounts, which looked like they were sent by employees of real company in Taiwan.

According to CNBC, two co-conspirators incorporated a company that posed as another company, Taiwan-based Quanta Computer – which actually does business with Facebook and Google. Then they created convincing forged emails using fake email accounts, which looked like they were sent by employees of the actual Quanta in Taiwan. They sent phishing emails with fake invoices to employees at Facebook and Google who "regularly conducted multimillion-dollar transactions" with Quanta, and those employees responded by paying out more than $100 million to the fake company's bank accounts [9].

8 Endpoint Protection

8.1 INTRODUCTION

This chapter is devoted to endpoint protection with the knowledge that there is some overlapping content with virus detection, which we discussed in Chapter 6. Endpoint protection, or endpoint security, is the methodology that an organization takes to protect its network when it is accessed by remote devices such as smartphones, laptops, tablets, or other such wireless devices. Endpoint protection technologies are software programs which are installed on all network servers and on all devices, which are at the endpoint. With the rapid increase in the number of mobile devices like notebooks, smartphones, laptops, etc., there has been a drastic increase in the number of such devices that are lost or stolen.

These lost or stolen devices can be a massive loss of sensitive data for enterprises. They have to secure and take care of their enterprise data that is available on their employees' personal devices. This would protect against their data not getting compromised, should those personal devices fall into the wrong hands. Hence, enterprises end up securing their endpoints, and this process is known as endpoint security. As such, endpoint protection technologies target endpoint security and also prevent the misuse of data on an employee's mobile device or an employee purposefully trying to cause harm to the enterprise.

While the objective of endpoint protection technologies is the same as antivirus technologies, there is a considerable difference between the two. Antivirus is about protecting just the computer or specific device, whereas endpoint protection technologies cover the big picture and secure just about every aspect of the network. Endpoint security involves different things like network access control, endpoint response and endpoint detection, application whitelisting, and most things that are not available in an antivirus package. Endpoint protection technologies can also be broadly classified into two different types of endpoint security:

1. For consumers who do not have any centralized management and administration
2. For enterprises where centralized management is absolutely necessary

Endpoint protection technologies have given rise to the endpoint protection platform (EPP). EPP prevents file-based malware, detects and blocks malicious activity from various applications, and investigates incidents or alerts while also working toward finding remediation. EPP's capabilities also include prevention, endpoint controls, detection and response, managed services, etc.

DOI: 10.1201/9781003038429-8

This chapter will review various endpoint protection technologies, explain how the technology works, the advantages and disadvantages of using this technology, and the different products which use end point technologies.

8.2 BRIEF HISTORICAL BACKGROUND

From the early 1990s, as use of the Internet began to rise, so came the increase of global viruses. As enterprises started to expand, the number of endpoint devices used to access enterprise data and enterprise networks began to grow. Nowadays, there are many mobile devices such as laptops, tablets, smartphones, etc. using these networks, and employees use these devices to carry out their daily job activities. When these devices get lost or stolen, there is a considerable risk for the enterprise's sensitive data falling into the wrong hands.

To address this inherent risk that comes with lost or stolen devices, enterprises started securing their data on these devices so that it was always protected, as well as securing access to the enterprise network.

Endpoint security technologies have gradually evolved into the following four major categories [1]:

8.2.1 TRADITIONAL ANTIVIRUS

Traditional antivirus technology dates back to the 1980s. Antivirus software offers support only for detecting malware and not for other advanced threats. To compensate for this shortcoming, antivirus technology had to be coupled with endpoint security, which led to the rise of different technologies.

8.2.2 ENDPOINT DETECTION AND RESPONSE (EDR)

The next stage in the evolution of endpoint security was EDR software, which succeeded in supplementing the shortcomings of antivirus software and offered much more reliable protection against security threats.

8.2.3 NEXT-GENERATION ANTIVIRUS SOFTWARE (NGAV)

Unlike traditional antivirus software, modern NGAV detects malware using machine learning and artificial intelligence techniques. Although it shows improved detection efficiency as compared to its conventional counterpart, it still needs improvement in order to be 100% efficient.

8.2.4 OPERATING SYSTEM (OS) CENTRIC SECURITY

Security rules are applied at the operating system level. Rules are formed according to what an application should or should not do and stored at the OS level. Tools such as SELinux, AppArmor, etc. enforce these rules and protect the system from data exfiltration, encryption, file manipulation, etc.

8.3   HOW ENDPOINT PROTECTION TECHNOLOGIES WORK

ANALOGY

An endpoint is like a home or apartment where all outside entry points or communication channels such as visitors, phone calls, packages received, or any other exchanges are carefully monitored to ensure no dangerous or unauthorized packages are delivered (Figure 8.1).

FIGURE 8.1   Endpoint protection.

8.3.1 ANTIVIRUS

Since antivirus was covered extensively in Chapter 6, we will simply provide a brief review here. Antivirus software has evolved and expanded in scope over the years, in order to combat ever changing threats. Today's best anti-malware solutions use different tactics to help protect your PC and MacOS desktops, as well as smart devices and networks.

There are three main methods of antivirus software that are most commonly used.

8.3.1.1 Signature

The most used method of all. After a virus has been identified on a computer or device, it is stored in a signature database locally or on a cloud, where it can be updated frequently. Every new file added to the computer goes through this list of viruses, and if any file matches, it is deleted.

8.3.1.2 Behavioral Detection

Instead of relying on files in the software, behavioral detection antivirus software keeps track of the behavior of the program. If it sees any abnormal or malicious activity, it raises the alert and removes the software.

8.3.1.3 Machine Learning

Analyzes the code of the application and predicts if it is malicious or not. The model keeps on learning from instances and keeps on getting better with an appropriate type of learning algorithm and a suitable amount of training data.

8.3.2 SECURE EMAIL GATEWAYS (SEGs)

Security email gateways reroute inbound and outbound email via proxy through its Mail Transport Agent (MTA), which then performs an email scan. The scan looks at different aspects of the email to decide whether it contains threats. If so, it filters the email.

Dynamic security email gateways also determine which emails are pernicious. These include flagged keywords, blacklisted URLs, or other suspicious qualities that suggest an email may contain a security threat. Artificial intelligence-driven applications are also used to prevent fraud emails from growing.

8.4 ENDPOINT PROTECTION TECHNOLOGIES

Endpoint security is a very diverse field and can comprise a variety of specific technologies. Three widely used endpoint technologies are reviewed below.

8.4.1 ANTIVIRUS SOLUTIONS

Antivirus software, covered in Chapter 6, is primarily used to prevent, detect, and remove malware. Originally, antivirus solutions were developed to detect and remove computer viruses. But, with the increase of other types of malwares, antivirus solutions started taking care of other security threats, such as ransomware, worms, keyloggers, spyware, etc.

Antivirus software often employs some of the following techniques to detect malware:

- Heuristics: Instead of a single signature, antivirus software often detects a generic signature that belongs to a family of viruses.
- Sandbox detection: The malware is first tested in a virtual environment to observe how it behaves, and if everything is okay, it is declared okay to be executed in the normal environment.
- Data mining techniques: A more modern approach where data mining and machine learning techniques are used to classify behaviors of an incoming program as malicious or safe.

Antivirus software is one of the most widely known and popular types of endpoint security solutions. Enterprises install antivirus software directly on their endpoint devices to protect them from malware. However, antivirus software often lacks capabilities when it comes to defending against advanced cyber threats. Many times, enterprises often think that installing antivirus software is enough to keep their endpoints secure, which leads to an increased risk of cyber-attacks.

8.4.2 Endpoint Detection and Response

Another endpoint security technology is EDR that pulls data from the enterprise's endpoints, analyzes it to reveal threat information. The data is stored in a central database. EDR software is directly installed on the endpoints and continuously monitored.

As compared to antivirus software, whose main job is detecting malware and preventing it from staying on an endpoint device, EDR solutions often detect malware and threats that escape from antivirus software. EDR immediately detects malware and prevents it from moving laterally in an enterprise network [2].

EDR solutions offer granular visibility, behavioral protection by studying indicators of attack, and a threat database that can be maintained for data analysis purposes. EDR solutions are often cloud-based solutions, which further add a layer of convenience for enterprises. Whenever an EDR software identifies potential threats to devices, it sends fast responses to endpoint users with alerts and ways to mitigate the threat.

8.4.3 Secure Email Gateways

Email is the most common mode of communication and hence a popular choice for attackers to threaten enterprise security. Hackers often use emails as a vehicle to carry out their attack for malicious attachments, executables, phishing emails, and ransomware. Through the years, methods of attack through emails have become more sophisticated, targeted, and increasingly threatening. Thus, it is necessary for enterprises to set up SEGs in order to mitigate attacks. SEGs filter emails successfully to protect sensitive enterprise data and user credentials.

SEGs perform various functions such as:

- Virus Malware Detection: SEGs scan emails regularly to detect threat patterns and email attachments to detect and quarantine those with malware or threats.

- Spam filtering: One of the functions of an email gateway is spam filtering. Spam is filtered using many techniques such as pattern matching, keyword analysis, etc. Spam emails are flagged and stored in a dedicated folder.
- Content filtering: Applied to outbound emails to filter the content being sent out from an enterprise. It is also used to check whether emails containing sensitive enterprise data are being sent.
- Email archiving: Helps with compliance by reducing the amount of emails being exchanged each day, which for many enterprises is a significant number.

8.5 ADVANTAGES AND DISADVANTAGES OF ENDPOINT PROTECTION TECHNOLOGIES

8.5.1 ANTIVIRUS SOLUTIONS

8.5.1.1 Advantages

- Provide protection against spyware, malware, rootkits, keyloggers, viruses, trojans, and worms [3].
- The firewall feature of antivirus software blocks incoming suspicious connections to a user's system. A firewall prevents hackers from infiltrating the system and stealing information.
- Provide secure Internet surfing. Phishing attacks are only possible if a user clicks on a link. With secure Internet surfing, users can know when a system is getting redirected to an unsafe or untrusted website.

8.5.1.2 Disadvantages

- Because of the ubiquitousness of antivirus software, hackers have developed many ways to attack and circumvent antivirus technologies.
- Antivirus is one type of endpoint security technology, but many users and organizations consider it as the only endpoint solution, decreasing the evaluation of specific needs for security technologies.
- A firewall feature is not present in every antivirus product available in the market.
- Antivirus technology needs to perform scans and checks on a system periodically. This increases its CPU usage, slowing down the system or network.

8.5.2 ENDPOINT DETECTION AND RESPONSE

8.5.2.1 Advantages

- EDR detection detects an attack earlier than other similar technologies. Faster attack detection leads to quicker actions and less resulting damage.
- Upon detecting an attack on a system, EDR can give a complete story of the attack, including the first infiltrated system, exposed systems and data breached. The complete story of an attack can help to reduce damage.
- EDR can determine if a system is under attack by using security analytics as opposed to network security, where individual suspicious activity can only be identified.

8.5.2.2 Disadvantages
- EDR solutions are complex and expensive to implement.
- EDR solutions depend upon skilled cybersecurity professionals to identify and detect issues, in spite of the use of Artificial Intelligence and Machine Learning techniques.
- EDR provides no contextual data outside of endpoint data. The data that EDR provides does not specify sources like firewalls, and there is a greater need for deliberate human efforts to analyze the data.

8.5.3 SECURE EMAIL GATEWAYS

8.5.3.1 Advantages
- SEGs protect individuals in organizations from spam, phishing, viruses, and social engineering attacks.
- Many SEGs offer email archival and email encryption facilities. These help secure sensitive data and can store email copies for legal issues.
- Secure email gateway service providers can provide cloud-based email service in case the email service provider goes down. Secure email gateway service ensures no email downtime for organizations.
- They can detect when personal or proprietary data – such as social security number, health-care records, private details of deals, and credit card information – are transmitted outside of a network. Such emails can also be stopped before transmission. The technique used is data loss prevention.

8.5.3.2 Disadvantages
- Monitoring unencrypted emails can lead to privacy invasion of individuals using the service.
- For SEGs with email encryption facility, only the text within the email is secure, the header of the email containing the subject and the recipient are still seen. A lot of information can still be known with this.

8.6 WHAT PRODUCTS USE ENDPOINT PROTECTION

Different products use endpoint protection technologies to protect a variety of endpoints such as servers, desktops, mobile devices, and devices that are embedded, such as printers and routers. The challenge of individual endpoint protection or security products is how to manage all individual endpoints effectively.

There are three main endpoint protection technologies.

8.6.1 ANTIVIRUS SOLUTIONS

Antiviruses are installed on our laptops, desktops that are used daily in a lot of organizations, desktops installed in schools and colleges, and nowadays even mobile devices have their own antivirus solution in the form of an antivirus software package. The most common ones are McAfee, Trend Micro, Bitdfender, and Kaspersky which have different types of subscriptions and packages.

8.6.2 ENDPOINT DETECTION AND RESPONSE

According to Gartner, EDR revenues more than doubled in 2016 and amounted to almost $500 million [4]. The major vendors that are responsible for more than half of this amount are Carbon Black, CrowdStrike, Tanium, and FireEye. Amongst others accounting for this revenue are Cisco, Symantec, RSA, etc. This tells us that almost every device which has a vulnerability of endpoint security compromise is using EDR to reduce its risk.

The main products that use this technology are:

- FireEye Endpoint Security
- Carbon Black Cb Response
- Symantec Endpoint Protection
- RSA NetWitness Endpoint
- Cisco Advanced Malware Protection for Endpoints
- Tanium

8.6.3 SECURE EMAIL GATEWAYS

SEG can be considered as firewalls for emails. It is very important to secure the server through which every email within an organization goes, whether it is incoming or outgoing.

Therefore, a lot of organizations now use SEGs. The main products currently using this technology in the market are:

- Proofpoint Essentials
- Barracuda Essentials
- Cisco Cloud Email Security
- Microsoft Advanced Threat Protection
- Forcepoint Email Security

8.7 CONCLUSION

According to IDC industry research, 70% of successful data breaches occur through endpoints [5]. Organizations need endpoint protection technologies to secure end devices like laptops, mobiles, computers. Considering that humans are the weakest link in the security chain, the second weakest link are the endpoint devices that are employed by users in their everyday tasks. Therefore, protecting endpoint devices goes a long way in providing greater security for organizations.

While network-based security technologies focus on a particular attack by checking suspicious activity through the network, the use of endpoint technologies helps secure data and predict if a system or an organization is under attack. Using endpoint technologies, the compromised data can be detected.

Most organizations assume that endpoint protection can be achieved by simply installing antivirus software on each system. However, an antivirus solution is one part of endpoint protection technologies that need to be used in combination with other technologies. Because of the popularity of antivirus solutions, nearly all organizations/individual users have antivirus installed. This has led to the development of hackers being able to successfully bypass antivirus techniques.

EDR solutions work as a complete system to detect an attack and take remedial measures. EDR further gives the complete story of an attack.

SEGs ensure a stop to spam, phishing, social engineering attacks. SEG providers also provide email encryption and archiving. SEGs make endpoint protection better by showing threat-free emails to users and ensuring that sensitive data does not go out of network.

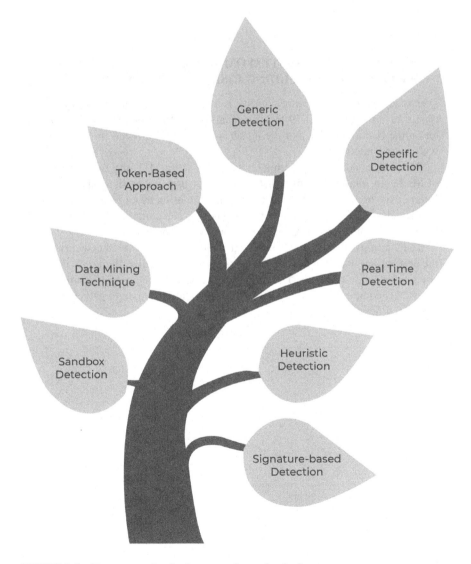

FIGURE 8.2 Taxonomy of endpoint protection technologies.

The use of Artificial Intelligence (AI) and Machine Learning techniques has significantly improved the performance of endpoint protection technologies. However, skilled human oversight is needed to avoid blunders in the detection of attacks and suspicious activities. Endpoint protection technologies are launched as complete software. The technologies can use the Software as a Service (SaaS) to switch to cloud for organizations. This ensures that all devices can access the endpoint protection software by connecting to the cloud. The use of the cloud ensures data security, scalability, and lower costs.

While security attacks are unavoidable, the use of proper endpoint protection technologies reduces the attack time significantly. Figure 8.2 shows the taxonomy of endpoint protection technologies.

ONLINE VOTING SYSTEM OMNIBALLOT HAS SEVERAL WEAKNESSES, INCLUDING RISK OF ENDPOINT ATTACKS

Discussion over online voting security is a rampant one that is more relevant than ever during COVID-19 lockdowns all around the world. OmniBallot is an online ballot delivery and voting service currently utilized in three states within the United States.

Research conducted by computer scientists from MIT and the University of Michigan found: "We find that OmniBallot uses a simplistic approach to Internet voting that is vulnerable to vote manipulation by malware on the voter's device and by insiders or other attackers who can compromise Democracy Live (online voting system), Amazon, Google, or Cloudflare."

The findings of the study don't find major concerns over the security of the application itself – but rather attacks that would utilize the user's device (endpoint) or side-channels (Democracy Live, Amazon, Google, Cloudflare) If a device is compromised, there is little the application can do to prevent abuse from an attacker [6, 7].

9 Malware Protection Technology

9.1 INTRODUCTION

With the introduction and rapid proliferation of smart technologies, it seems that many things became "smart": smart phones, smart cities, smart environment, etc. Since smart devices are only increasing in use and popularity, they are also becoming a more common target for cybercriminals, who are finding new ways to target them with threats. Malware – which is defined as malicious software created or used to disrupt computer operations, gather sensitive information, or gain access to private computer systems – continues to be a significant threat for smart devices and the need to provide security against malware is absolutely essential.

The most common malware are viruses (see Chapter 6) and while they are a type of malware, they are technically different [1]. Viruses are malicious codes that are attached to a program and target a user, while malware is itself an application working independently if installed on a system.

Worms are another kind of malware that begin when one machine gets infected, then spread throughout a network and infect all the computers that are connected to it. Spyware is another type of malware, which spies on users and collects their private information (i.e., credit card number) without the user's knowledge. Another malware is Trojan, which behaves like a normal program but creates a hidden door to breach the security in a system. Trojan allows other malware to have easy access to a system. Lastly, there is ransomware that encrypts all the information and files on a user's computer and asks the targeted user(s) for money to retrieve their files and data. Ransomware is an especially difficult challenge since a targeted user might pay the requested ransom, but not get their encrypted files back. Or they may be attacked again because they demonstrated vulnerability and a readiness to pay.

Since there are numerous kinds of cyberattacks, as many technological solutions have been developed to protect user devices, including many anti-malware programs.

This chapter will review the various technologies used in combating malware, how the technologies work, advantages and disadvantages of each, and the different products that use these technologies.

9.2 BRIEF HISTORICAL BACKGROUND

Malware is a malicious software. In its early age, malware was very primitive and began by spreading entirely offline, via floppy disks carried from computer to computer by human hands. With advances in networking and the Internet, so too has malware also become more sophisticated and nuanced in the threats it poses.

DOI: 10.1201/9781003038429-9

Since 2000, malware has significantly grown by the number and speed of infections it has spread. Well-known malware attacks – such as the Sony DRM Rootkit in 2005 (music CDs that secretly installed a rootkit on computers), or Crimeware (a variety of techniques to steal confidential data) or SQL injection (a code injection technique that destroyed databases) – brought into focus the need to find technological solutions to malware attacks.

In more recent times, some of the most famous malware attacks have included the Stuxnet Worm (2010), Backoff (2014), or WannaCry Ransomware (2017), to name just a few. Malware has expanded from merely infecting computers to attacking virtually anything with a microprocessor. Malware targets now are not limited to PCs, but can infect wearables (like watches), light bulbs, automobiles, water supply systems, and even airliners [2].

Historically, the basic method to detect malware was to check if the signature existed in a database, a methodology known as signature-based detection. This process figured out the presence of a malware attack by comparing at least one-byte code sample of the software in query with a database of signatures of known malicious packages, known as blacklists. A signature-based detection scheme was based on the assumption that malware can be explained or justified through different patterns (also known as signatures). Signature-based detection has become the most commonly used approach for anti-malware systems. But this method was limited as it only detected malware that already existed in the database and not newer versions.

Detection evolved to be able to find new malware by analyzing the program, its behavior, and its structure. Heuristic analysis (HA) technology started at the end of 1987. At that time, there were two primary malware detectors being used, *Flu Shot Plus* and Anti4us. These functioned differently from how detectors are working today, but the goal was simply to protect computers from malware. They divided the binary numbers into a code section and data section. Then, as technology improved, a more advanced HA aspect was added [3].

The host-based intrusion prevention system (HIPS) is a variation of the Intrusion Detection System (IDS). The first real-time model for this was researched and developed in 1984 and 1986. The prototype was named an Intrusion Detection Expert System (IDES). It was a rule-based system and was only trained to detect known malicious activities [4]. In the ensuing year this prototype evolved into the following variations:

- Host-based: In this method, the focus was on a single host machine. The data from the machine was analyzed to detect signs of an attack as the instruction packets entered or exited the host.
- Network-based: Memory was used to record suspicious data. The network's incoming data was scrutinized against a database and flagged those that looked suspicious. Audit data could be collected from one or several hosts and added to the database.
- Anomaly detection model: This system had knowledge of expected behavior (schema) and depended on being able to detect any outliers or deviations from the established baseline. Since it found any deviations to be suspicious,

it had a high count of false positives. However, the same trait enabled it to detect unknown intrusions.

• Misuse detection model: In contrast to the previous case, this system had knowledge of suspicious behavior and search activities that violated the established policies. It also looked for known malicious or unwanted behavior. The advantages were its efficiency and high accuracy rate (i.e., low false positive count).

The next phase in IDS came in the early 2010s, with the incorporation of MD5/SHA checks and the inclusion of features such as application and user controls [4].

9.3 HOW MALWARE PROTECTION TECHNOLOGIES WORK

ANALOGY

Malware detection functions like an airport security checkpoint. All packages, including what a traveler has in his clothes, are carefully scanned for all prohibited and dangerous objects and any such objects are confiscated if found. There are objects that are easier to find – such as a gun – but there are plenty of other materials that might be harder to see and would need a closer inspection. There is also the challenge of encountering objects that are unknown or whose status is unclear to inspectors. Thus, like airport security, malware detection methods have a variety of threats to assess and prevent (Figure 9.1).

FIGURE 9.1 Malware detection.

9.3.1   HEURISTIC ANALYSIS

HA is used in most antivirus and malware solutions. It is a technology that analyzes the structure of the program, its behavior, and other features to be able to detect different kinds of malware even if this malware is new and does not have similar features as the malware in the database.

When a heuristic system starts scanning the files in a system, mainly the executable file, it inspects data, program logic, and the overall framework. Then it searches for unusual behavior, structures, instructions, or suspicious code, which could appear as junk code. Next, it determines if there is malware based on the evaluation of the behavior and overall structure of the program. It uses different scanning techniques such as genetic signature detection, file analysis, and file emulation. In genetic signature detection, it uses previous malware signatures to discover new malware with the same family. In file analysis, it examines the software closely to verify its purpose, intent, and destination. If the software behaves suspiciously, it marks it as dangerous. For file emulation, which is known as a sandbox testing, it tests the file in a virtual environment and watches its actions. If the software acts strangely, it marks it as unsafe.

9.3.2   SIGNATURE-BASED SAFETY

While detecting, all devices have significant parameters that may be used to create a unique signature for individual devices. Algorithms can be a little more advanced to effectively test an item to determine its digital signature.

Whenever an anti-malware solution issuer finds an item as malicious, its signature is brought to a database of acknowledged malware, which might contain a large number of signatures that understand malicious items. This technique of figuring out malicious objects has been the primary method utilized by malware products and utilized by current firewalls, electronic mails, and network gateways.

Signature-based malware detection technology has some strengths. It is, without a doubt, the most widely known and understood, and the first actual antivirus packages used this method. It is likewise fast and smooth to run. Above all else, it provides protection from the thousands and thousands of older malware attacks.

9.3.3   HOST-BASED INTRUSION PREVENTION SYSTEMS

HIPS are used mostly in conjunction with a detection system like the ones mentioned above. Once a suspicious activity is reported, HIPS can conduct a variety of actions to shut down the threat, which include: restoring the log files available in the storage, suspending suspicious user accounts, blocking suspicious IP addresses, killing compromised processes, shutting down corrupted systems, starting up processes to help scan or clean the device, updating firewall settings like adding websites to be blacklisted/whitelisted and alerting, recording, and reporting suspicious activities to authorities.

9.4   MALWARE PROTECTION TECHNOLOGIES

There are two main components in technology protection: technical and analytical components. The technical components are defined as a group of collected programmed algorithms and functions which gather inputs, dismember them, and send them to the next phase to be analyzed. The inputs can be a file byte code, text strings in a file or a sequence of different actions. The analytical component is a decision-making method. It consists of a group of algorithms that investigates the data that was received in the previous phase, and then draws a conclusion based on that analysis. The antivirus program will then take action based on the decision that was made by both phases [5].

Figure 9.2 shows the main category of malware detection technologies. To act against the ever-evolving range of risks, anti-malware programs need to provide many layers of safety.

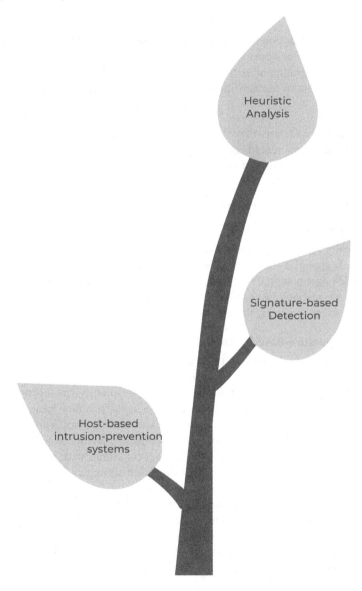

FIGURE 9.2 Taxonomy of malware protection technologies.

The initial layer comes from a database of blacklisted URLs and IP addresses with updated information in real time that block those URLs. Next, a report without a stated URL or IP is classified as an accurate file, or one requiring similar inspection. Then, heuristic- or conduct-based layers decide whether to prevent a document from executing the code, based completely on its supposed action. If an action is uncommon or dangerous, this aspect of malware safety will categorize it as malicious. Finally, a technique called sandboxing is frequently used to isolate a report when there aren't enough records to make a ruling on a document's category. Many anti-malware solutions offer sandboxing, but the speed and effectiveness with which this procedure is performed ranges substantially.

Only a small fraction of malware threats makes it past the first layer of strong malware safety. But because of the sheer volume of threats getting through to online users, a fragment still represents a critical danger. According to some estimate, nine out of ten net-linked personal computers are infected with adware that can:

- Be used for identity theft
- Expose private records and personal bills
- Capture passwords and usernames

9.4.1 MAIN CATEGORY OF MALWARE DETECTION TECHNOLOGY

9.4.1.1 Heuristic Analysis

HA basically searches for commands and instructions that appear to be unusual in a program. It applies a weight-based system to evaluate a program and determine the magnitude of the potential threat. If it exceeds a predetermined threshold, the antivirus using this technology can take action, such as sending an alert to the administrator of the server or putting the file in quarantine.

9.4.1.2 Signature-Based Security

All malware has a completely unique signature (a unique string of bits, cryptic hash, or a binary pattern) that may be taken into consideration as a fingerprint for its identity. Since the inception of malware, most antivirus technologies have been using signature-based malware detection as the primary weapon in opposition to malware-laden intrusion attempts. Anti-malware software might reveal all of the data getting into a system and scan the contents to test if the source code or hashes inside the documents or packets match with any of the known malware threats. Signature-based total detection strategies have been simple to enforce and replace for vendors.

Consequently, all anti-malware companies maintain their library of recognized threats. The efficiency and accuracy of these software programs have been measured by assessing which have the most malware signatures and their abilities to include new signatures and push them to customer systems. This technique provides dependable protection against millions of acknowledged and active threats.

9.4.1.3 Host-Based Intrusion-Prevention Systems (HIPS)

Since there is a large variety of malware and they are constantly changing their signatures, HIPS focus on the attributes of the malware and its behavior. They monitor a single host for suspicious activities by analyzing the events occurring within it. They study the behavior of the code in order to stop the malware. They protect the computer against known and unknown malicious attacks by blocking any major changes attempted by the malware and alerting the user so an appropriate action can be taken.

9.5 ADVANTAGES AND DISADVANTAGES OF MALWARE PROTECTION

9.5.1 ADVANTAGES

9.5.1.1 Heuristic Analysis

HA detects malware in boot records and files by analyzing its characteristics before the malware runs in the system and in web pages. If it recognizes something suspicious, it blocks the website. HA can detect malware that has a different signature than the one that was stored in the database. It is also faster than other techniques like sandboxing.

9.5.1.2 Signature-Based Protection

Signature-based protection is a practical approach that focuses on specific attacks and is very accurate at lowering the rate of malware attacks. Signature-based protection technology compares hashes and signatures of files to a list of recognized malicious documents. It also looks inside documents to discover signatures (defined as a fingerprint of a given attack) of malicious code. Signature-based protection only detects fingerprint patterns and if it does not identify a pattern, it indicates a failure. Thus, it does not work without a pattern identification.

9.5.1.3 Host-Based Intrusion Prevention Systems

HIPS (Host Intrusion Prevention System) is effective at blocking intruders. When an intruder begins conducting activities on a network, at this time only, it starts taking measures. It allows the user or application to implement their preferred plan of action. This gives a level of customization since the user can define actions corresponding to certain attacks.

9.5.2 DISADVANTAGES

9.5.2.1 Heuristic Analysis

HA does not give detailed information about how malware is flagged.

When scanning a sample, the information usually is limited to the threat name.

HA might slow down the performance of the system, and it can report false positives, meaning that a safe program could be marked as dangerous. It may also be limited in its ability to detect new malware.

9.5.2.2 Signature-Based Protection

Signature-based protection products are designed to only search for hits and protect against recognized attacks. As such, they cannot detect an attack for which there is no corresponding signature that was previously stored in the repository. Thus, it cannot shield a website from new or unknown enemies.

Signature-based protection only detects fingerprint patterns. If it does not identify any such pattern, then it indicates a failure. A major disadvantage is that it can result in false positives, i.e., marking safe programs as malicious. Additionally, for users to make a valid decision, they need to have at least a decent knowledge of computing.

9.5.2.3 Host-Based Intrusion Prevention Systems

HIPS is not designed as a solution for all potential threats and does not include software patches that manage mentor configuration controls for network devices.

9.6 WHAT PRODUCTS USE MALWARE PROTECTION

Some of the main products that use each type of malware detection technologies are:

- HA: Malwarebytes, McCafe, Sophos, Bitdefender, avast
- Signature-based protection: Malwarebytes, McAfee, ScanGuard, TOTAL AV
- HIPS: McCafe, Splunk, Fail2ban

9.7 CONCLUSION

Malware is a collective name for a wide range of malicious software like spyware, ransomware, Trojans, and worms, to name a few. Since malware is constantly modifying and adapting to breach the security of devices and networks, anti-malware technologies, therefore, must now be able to defend against known and unknown threats.

Several technologies are used in anti-malware products such as HA, signature-based analysis, and HIPS. The technologies explained in the above sections are a few of the many available and are used in conjunction with others to form effective anti-malware. Future work will involve artificial intelligence and machine learning techniques to be able to detect and protect devices. Figure 9.2 shows the taxonomy of malware protection technologies.

RESEARCHERS DISCOVER MALWARE ATTACKING PHYSICAL INFRASTRUCTURE

Between 2015 and 2016, several power outages occurred on the Ukrainian power grid due to an attack believed to be caused by Russian attackers.

The 2015 malware attacks were accomplished with a more manual approach of switching off power to electrical substations, but in 2016 the attacks were more advanced and automated, which meant that they occurred more quickly, with less preparation and with little human involvement. In addition to the improved efficiency of the malware, its ability to scale was equally better than the early attacks.

As intimidating as this story might sound, detecting malware in advance may be a simple solution. According to Wired, cybersecurity firms ESET and Dragos, Inc., Crash Override (a crisis helpline, advocacy group and resource center for people who are experiencing online abuse) can be used to affect other types of critical infrastructure such as transportation, gas lines, or water facilities.

This example proves the importance of increasing and improving malware detection methods. Being able to respond to incidents before damage is carried out is critical to maintaining not only the security of devices but also, potentially, the critical components of infrastructure and society [6].

THE SOLARWINDS HACK

The SolarWinds hack, one of the largest and most sophisticated attacks ever, happened in December 2020. Hackers secretly broke into SolarWinds systems and added malicious code into the company's Orion business software updates system. When Solarwinds, a major US information technology firm, sent out updates for bug fixes or new features to 33,000 customers, it included hacked code. Multiple US government agencies including the State Department, Treasury, and Homeland Security were targeted in the attack [7, 8].

10 The Internet of Things (IoT)

10.1 INTRODUCTION

We live in a world of expanding connectivity. Not only are we as people able to connect with each other at greater distances through various means, but our devices are capable of connecting with each other in a similar manner as well. We refer to these growing systems of interconnected devices – which communicate with and transmit data to each other through networks without human to computer interaction – as "The Internet of Things" (IoT). IoT is growing rapidly; every device is connected to the world through the Internet and hence security is a major point of focus. According to Gartner, 8.4 billion connected "Things" were in use in 2017, up 31% from 2016 [1, 2].

The number of IoT devices is exponentially rising in all areas from hospitals to manufacturing to workplaces to homes. These interconnected devices access a broad range of information about users that can be personal and sensitive. The progress in machine learning and artificial intelligence makes all these devices even more interactive and efficient.

Security is paramount for these devices. Unsecure devices can lead to devices being hacked, hijacked, and potentially severe damages. To prevent these devices from being hacked, it is important to have essential security protocols in place within them. IoT devices are vulnerable and can be hacked and utilized for larger, more sophisticated attacks. This capacity provides means for which malicious actors who target and hack IoT devices could gain control of personal information and even control the physical surroundings of users. There are numerous examples of hackers exploiting security vulnerabilities in IoT devices.

Hacks can be as surprising as using an IoT fish tank installed in a casino to access the casino's network and send out 10GB of data [3]. And hacks can be as sophisticated as hijacking hundreds of thousands of IoT devices to attack a company server which serviced major websites such as Twitter, Spotify, Netflix, and the New York Times [4]. The potential for personal targeted attacks is clear when a high-school student hacked about 150,000 IoT printers to print messages informing owners of the hack [5]. Although it is harder to find specific reports of targeted personalized attacks, data shows that 67% of people who are victims of a data breach are also victims of fraud [6]. All of these examples merely scrape the surface of the constant utilized potential of IoT security vulnerabilities in causing disruption and real harm to individuals, companies, and infrastructures.

Many important security features are essential to securing IoT devices. Features like the discovery of devices, encryption of data, authentication, brute force protection, and Web API security (transfer of data through APIs (application programming interfaces) that are connected to the Internet) are some of the most important ones.

DOI: 10.1201/9781003038429-10

Although IoT devices are, in many ways, like normal devices, they are connected to the Internet, which makes identifying them an especially difficult task.

This chapter will review the basic IoT technologies from a security perspective, explain how the technologies work, and the advantages and disadvantages of IoT technologies.

10.2   BRIEF HISTORICAL BACKGROUND

In the early days of computers, the concept of interconnecting all computer devices to each other was developed and shaped with the Advanced Research Projects Agency Network (ARPANET) in 1969.

In 1990 John Romkey created a toaster that could be turned on and off over the Internet. The toaster was connected to a computer with TCP/IP networking [7].

IoT, a system of interrelated computing devices terminology, was described by Ashton in 1999 and started to be used. In 2003, The Guardian defined the IoT as "the tiny microchip that measures less than a third of a millimeter wide, little bigger than a grain of sand. It can contain information on anything from retail prices to washing instructions to your medical records" [3].

From 2009, IoT became a recognized area. In 2011 Internet Protocol version 6 (IPv6) was publicly launched, allowing more unique TCP/IP address identifiers to be created. IPv6 provided IoT products a platform to operate on for a very long time, thus allowing the connection of billions of devices to the Internet [8].

10.3   HOW IoT TECHNOLOGIES WORK

ANALOGY

To better understand IoT devices, consider this analogy. Imagine all IoT devices in your system are people invited to join a virtual group but that all communication must stay within the group and not get to people outside the group. Before joining your group, each person should first prove their identity and authenticate who they are. You protect your system so that no communication by chats, email, or any other means can get outside of this circle. You also will monitor and moderate the intergroup and block whatever you think should not be communicated.

In an IoT device, each device will ask the others to establish the device's identity and make sure it belongs to their network. Then the device will request communication from the other device. For example, a smart refrigerator contacts the smart thermometer and asks about the weather outside and inside. The smart thermometer first verifies the identification of the refrigerator and then answers the question (Figure 10.1).

IoT devices are Internet-enabled devices with an IP address that collect data and transmit it to a gathering point. Moving the data is done using a range of wireless technologies or on wired networks. The data is then sent over the Internet to a data

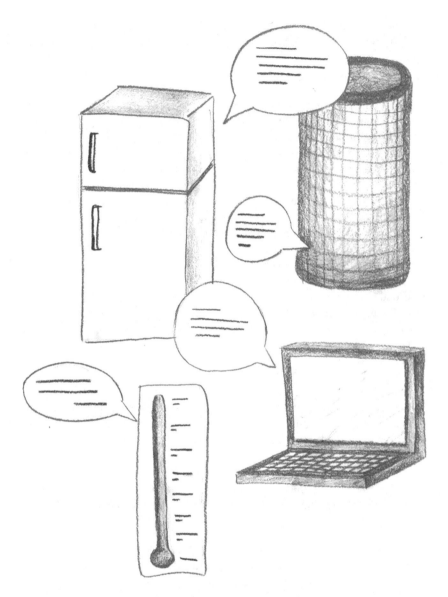

FIGURE 10.1   IoT devices communication to each other.

center or a cloud service to be processed, filtered, and analyzed, each of which can be handled in a variety of ways, with different technologies and on different platforms.

This process is the same for smart personal devices such as wearables, smart automobiles, smart homes, smart manufacturing systems, or smart cities.

An IoT platform connects edge hardware (that is entry point to network), access points, and data networks to other parts of the value chain and handles ongoing management tasks and data visualization.

IoT Platform types can be classified under End-to-End, Connectivity, Cloud, and Data Platforms.

10.3.1 END-TO-END PLATFORM

An end-to-end platform is used by businesses or service providers as a completely vertically integrated service offering or product architecture. For instance, Samsung offers an IoT platform to integrate all data sources.

An end-to-end platform offers all parts of the product architecture: (a) Hardware or Sensor Product, (b) Gateway or Edge Collector, (c) Mobile App as an Edge device, (d) Middleware component, (e) Message Bus or Message Queue injection that allow different systems to communicate through a shared set of interfaces, (f) Computing or Analytics platform, and (g) UI or Insights recommendations.

An end-to-end platform provides several benefits:

- Over the air Firmware (OTA) Upgrade
- Device Management
- Cloud Connection
- Cellular Connectivity
- Analytics
- User Interface

10.3.2 CONNECTIVITY PLATFORM

Connectivity IoT platforms enable direct integration of sensors via an edge device or gateway to the Cloud. Sensors upload the collected data periodically to the on-site collector device. Based upon the availability of network connections, the collector uploads to the cloud.

A connectivity platform potentially allows IoT devices to run for years on small batteries, occasionally sending out small packets of data, waiting for a short time for response messages, and then closing the connection until more data needs to be sent. The process can be summarized as follows:

1. Sensors communicate to a gateway through Near Field Communication, M Telemetry Transport (MQTT) over User Datagram Protocol (UDP), or LoRa.1.
2. The gateway, a processing system, assembles the data from the sensors and performs basic logic operations.
3. Using cellular networks or Wi-Fi, the gateway relays the data to the cloud.
4. The data is processed in the cloud and actionable recommendations are sent to the user.
5. Mobile recommendations are delivered.

10.3.3 CLOUD AND DATA PLATFORMS

IoT cloud and data platforms connect, store, and manage IoT data. They create a "meeting point" for all connected devices and serve to collect and handle the data they deliver over the network to the IoT cloud.

10.4 IoT SECURITY TECHNOLOGIES

IoT technologies provide connectivity to all devices, making them "smart" by connecting them to other devices in a network and "intelligent" by using machine learning and data science to manage and analyze real data in order to make decisions and predict future events.

IoT technologies include several technological layers: device sensors, gateway/communication, and application on the cloud. The interaction with a device is generally done through the user interface of the device App, if provided (some devices do not offer access to the device through a user interface).

Device sensors are hardware that collect data. The type of data collected can be very simple or very complex, and a device can have multiple sensors of each type. For example, a mobile phone might have an accelerometer, ambient temperature sensor location tracker (GPS), camera, and more.

A gateway or communication layer consists of sending collected data from sensors to the cloud through various mediums of communication and transports such as cellular networks, satellite networks, Wi-Fi, Bluetooth, wide-area networks (WAN), and low power wide area networks.

A cloud application consists of receiving, storing, and processing acquired data. It is important to highlight that there is a lot of data collected by each IoT device. Therefore, rather than sending all data to a centralized system, each device does some data processing first and sends only relevant material back to the cloud for more analysis. This is called edge computing.

An example of this could be a network of dozens of IoT security cameras. Instead of bombarding a building's security operations center (SoC) with simultaneous live-streams, edge computing systems can analyze the incoming videos and only alert the SoC when one of the cameras detects movement [9].

Interaction with a device might be available to users through an interface that communicates with the IoT system. For example, if a user has a camera installed in their house, they might want to check the video recordings and all the feeds through a web server.

Depending on the IoT application and complexity of the system, the user may also be able to perform an action that may backfire and affect the system. For example, if a user detects some changes in the refrigerator, the user can remotely adjust the temperature via their phone. The actions can also be performed automatically if there are predefined rules where the system might intervene without any involvement from users. The action might also be generating an alert not only to the users or any other third parties.

10.4.1 SECURING IoT DEVICES

IoT security protects Internet-enabled devices that connect to each other on networks. There are several levels of security technologies that are needed to protect IoT devices. The most important parts are authentication and tracking, data and information integrity, mutual trust, privacy and digital forgetting. Since a great deal of processing happens at a centralized location or in the cloud, there is a heightened need for ensuring security for the cloud [10].

IoT devices generally operate without batteries. Their energy is provided using a wireless medium. Many of these systems allow for only very minimal hardware and, thus, require a compact security solution with a small footprint and energy budget, since many IoT devices often operate in an unprotected environment. Therefore, the first level of security is ensuring that collected data is authenticated.

Some IoT sensors may have high bandwidth and low latency (real-time) data collection rates. Therefore, appropriate data integrity techniques such as encryption and watermarking are required.

It is likely that billions of devices will be a part of the IoT ecosystem. Each of these nodes should have a unique identifier. In addition, at any point in time, the IoT infrastructure should be able to track each item. Another level of difficulty is that many nodes will be placed in high density places and access to them may be hindered or even blocked. It is important to ensure that all collected data is authentic. Therefore, appropriate data integrity techniques such as encryption and watermarking are required.

There is an essential need for ensuring that each user can be guaranteed that the data presented by an IoT device is trusted, i.e., that it is indeed collected by the stated sensor at its stated location and at its stated time. Recently, several schemes have been proposed for ensuring IoT trust. These solutions are based on hardware security primitives – for example physical unclonable functions (PUFs) and random number generators (RNGs) – and should be further optimized in terms of cost and energy. Also, hardware and software attestation techniques may be used for trust-related tasks. Interestingly, another important problem is operator trust, in which sensors and IoT devices can authorize and trust the instructions of IoT users.

The most difficult privacy task is one where the attacker integrates information from different sets and modalities at the semantic level. Combining different data from different sources of information at the semantic level can result in the extraction of unexpected information [10, 11].

Data revocation (i.e., digital forgetting) is the process of deleting all copies of a data set [11]. In addition to the tremendous amount of sensory data that IoT devices will collect, there will be huge data sets related to communication activities between various users and IoT devices. It is plausible to expect that a significant percentage of this data will contain important information and knowledge about the users and their actions and interests.

There are several data revocation techniques proposed in classical cryptography. All of them are based on the simple idea that encrypted data is effectively deleted if the required decryption key is deleted. There are also several techniques that employ distributed data storage so that data is deleted due to unavoidable social and technical processes. These ideas are valuable and essential to IoT systems which can collect large amounts of data that can seriously impact the privacy of many individuals and even compromise the security of economic entities and government institutions.

A large percentage of devices will depend on harvested energy. In order to reduce energy consumption, computation tasks will often be offloaded to data centers. Communication will often use technologies that require less energy than those currently widely used. Most likely, near-field communication (NFC) will greatly increase its market share. Other highly constrained metrics include cost and area. It is likely that new packaging and integration technologies will emerge. It's important to note that many deployed devices should be in operation for years if not decades.

10.4.2 SMART DEVICE PROTECTION TECHNOLOGIES

10.4.2.1 Host Identity Protocol (HIP)

HIP is based on establishing a secure channel for signaling and support for mobility through a protocol layer. The technique of HIP does not completely remove public key cryptography as is followed in lightweight HIP. Instead, HIP is about the removal of the signature and using Elliptic-curve cryptography (ECC) which is a better key exchange method for sensor nodes.

10.4.2.2 HIP Diet EXchange

HIP **Diet EXchange** (DEX) is a variant of the HIP. It is a two-party cryptographic protocol used to establish secure communication between hosts. The first party is called the Initiator and the second party the Responder. The parties then authenticate each other.

The HIP DEX technique does not support Elliptic Curve Digital Signature Algorithm (ECDSA) based signatures and thus does not have an option for verification of remote identities. It uses some kind of external procedures for verification. The HIP DEX technique can also be used for a MAC layer in a Wireless Sensor Network (WSN) radio standard. It does not offer end-to-end security between sensors and base stations.

10.4.2.3 Intrusion Detection and Mitigation Framework

Intrusion Detection and Mitigation Framework (IDM) for IoT devices is a technique for securing smart devices at a network level. It monitors the activities of a smart device and searches for any malicious or suspicious activities. It also aims to stop intruders from accessing victims' devices. The goal of this technology is to offer services for security using software-defined networking (SDN) technology, but it is not based on a single machine learning technique. The IoT-IDM method observes the network traffic captured by the sensors of smart devices for identification purposes. It is also capable of raising an alert in an event of suspicious activity.

10.4.2.4 Intrusion Detection System

An Intrusion Detection System (IDS) is the process of monitoring the events in a computer network for possible attacks. There are two types of such systems, network-based and host-based IDS. A network-based IDS monitors network traffic and analyzes different network layers to identify possible threats and suspicious activity. A host-based IDS monitors suspicious activities of a single-host in a network. A host-based IDS will often deploy detection units – called agents – installed on target hosts. The sensor elements track the network activities of target systems in a typical computer network IDS.

Other components evaluate the information events obtained from these elements to indicate future incidents. There are two kinds of sensor elements, inline and passive. Inline sensor elements allow network traffic to pass through them while passive elements scan a copy of the actual network traffic.

An IDS will use three main techniques to investigate incidents: (1) signature-based, (2) anomaly-based, and (3) stateful protocol analysis. When a threat is recognized, then the pattern or signature corresponding to the threat or signature can be identified.

Signature-based detection is a process in which signatures are compared against observed network traffic to determine possible malicious activity. Anomaly-based detection is used where there is no threat detected beforehand. In this technique, in an effort to detect any malicious activity, the difference between the normal behavior is compared with observed network security. Therefore, it's important to establish a profile reflecting normal behavior. Finally, stateful protocol analysis depends on profound protocol operation inspection and monitors the status of application protocols used by the network or hosts. In reality, a typical behavior of a protocol is correlated with observed events in this system in order to detect anomalies.

10.4.2.5 OpenFlow

OpenFlow is a network protocol to manage and direct traffic among routers and switches from various vendors. It separates the programming of routers and switches from underlying hardware.

Switches/routers in computer networks have two main tasks: a control plane and a data plane. The control plane is responsible for making decisions on where network traffic should be directed, while the data plane forwards the traffic to respective destinations.

10.4.2.6 Software-Defined Networking

SDN centralizes network intelligence in one network component by disassociating the forwarding process of network packets (data plane) from the routing process (control plane). The control plane consists of one or more controllers, which are considered the brain of the SDN network and where the intelligence is incorporated.

10.5 ADVANTAGES AND DISADVANTAGES OF IoT SECURITY TECHNOLOGIES

IoT devices, from those in homes to those used in manufacturing to any type of SCADA (a centralized system that monitors and controls the entire area), facilitate communication among different devices, consequently making their respective network smart and intelligent.

10.5.1 ADVANTAGES

The advantages of IoT Security Technologies are as follows:

- Data Collection: Information is collected and can be used from any location
- Create communication among different devices on a network

- Make automation tasks easy with the help of machine learning and without human agent intervention
- Improve productivity by monitoring and automation
- IoT technologies are relatively inexpensive

10.5.2 DISADVANTAGES

The disadvantages include:

- High risk of leaking confidential information
- No standard protocols to create safe communications among devices
- A lack of privacy, resulting from data being transferred in the cloud or on a network
- Issues with one device might affect the whole network

10.6   WHAT PRODUCTS USE IoT TECHNOLOGIES

A smart bridge network and a wired connection that act as a bridge between smart devices and the Internet allow, connect, monitor, and control all smart devices from a single mobile device from anywhere.

Smart bridge networks and a wired connection to a network connect home networking to the outside world. A smart bridge serves as a connection between smart devices and the Internet. It monitors and controls all of the smart devices from a single mobile device. A wireless Zigbee protocol is used as a communication mechanism between smart devices, smart cameras, thermostats, smart plugs, security detectors, light bulbs, and smart street solutions. Entertainment and lock-based systems employ this technology.

10.7   CONCLUSION

IoT gives an amazing chance for controlling and having complete access to peer connected devices for any geographic location. Devices in the IoT also provide flexibility and reachability, as the connection can be made at any time and from anywhere. There are countless applications into which smart devices can be integrated, without the need for them to be wired or connected.

Although IoT connected devices are smart and have great benefits, they are also prone to some challenges. These come in the form of attacks that can be made on them due to a lack of security systems that monitor or safeguard them. Therefore, important and critical measures are required to mitigate and control the concerns of hacking these smart technologies. IoT-IDM frameworks help to find and mitigate those known attacks and methods that prevail among smart tech-oriented home environments. In order to accomplish this, the SDN architecture has been employed into the IoT-IDM. The architecture has taken the very advantage of visibility in networking and provided a very healthy and attack-proof solution for a secure environment for these smart devices.

Furthermore, IoT-IDM uses many machine learning-based techniques to find the hosts and hot spots that can be compromised and find the location from where the attacks are primarily launched. Once the source of the attack on an IoT-based system is detected, the IoT-IDM creates a set of policies. Once some of the important policies are generated, they are then moved over to dependent routers that switch networks to tackle the attacks against the smart device. Figure 10.2 shows the taxonomy of IoT technologies.

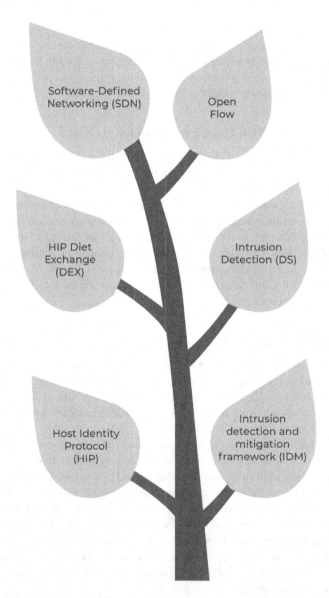

FIGURE 10.2   Taxonomy of IoT technologies.

FISH TANK ALLOWS A CASINO TO BE HACKED

In 2017, the Washington Post reported that hackers attempted to acquire data from a North American casino by using an Internet-connected fish tank within the casino. The fish tank had sensors connected to a PC that regulated the temperature, food, and the cleanliness of the tank. The hackers were able to access the fish tank, use it to move around the network, and send out data [12].

RESEARCHER UNCOVERED AN IMPORTANT FLAW IN SMART DEVICE INFRASTRUCTURES

According to a report by gizmodo.com, in 2020 a researcher discovered that their coffee maker acted as a Wi-Fi access point and used an unencrypted connection to link to its corresponding smartphone app. Users could then interact with their machine by connecting the coffee maker to their home Wi-Fi network. The app updated firmware through the home network without encryption or authentication, thus no code signing was needed. Consequently, the researcher uploaded the Android app's latest firmware version to a computer and reverse engineered it using IDA, an interactive disassembler, a staple in any reverse engineer's toolkit. The researcher then wrote a python script that mimicked the coffee maker's update process to implement the modified firmware and lines of script that actually triggered it to go haywire.

This example shows that with this type of exploit, attackers can render a smart gadget incapable of receiving future patches to fix a particular weakness. Then, attackers can program a coffee maker, or any smart appliance, with this vulnerability, to attack any device on the same network without ever raising any alarm bells [13].

11 Network Security

11.1 INTRODUCTION

Network security can be classified into a broad range which encompasses many different technologies, devices, and their interconnected processes. As the number of connected devices exponentially increases, the rules, standards, laws, and configurations for these different and unique devices must be defined for the way they communicate with each other. This is where network security comes into play. It protects the integrity, confidentiality, and data accessibility of computer networks. Network security is especially important in a time when cybercriminals are becoming more aware of exploiting new users, and enterprises that have a huge repository of public data must take extra precautions in protecting their servers from any type of data breach.

This chapter will review various types of network security technologies, explain how they work, the advantages and disadvantages of each, and the different products which use these technologies.

11.2 BRIEF HISTORICAL BACKGROUND

The evolution of network security technologies has occurred as a response to threats over time. The first-known penetration into a network occurred in 1967 [1], when students of a Chicago high school computer club had access to IBM's APL network, a programming language for users and developers. Originally, the APL programming was intended to work within its own APL workspace but after having learned the system, the students were able to hack into the system's wider context [2]. However, the intention was a simple curiosity, yet this penetration was later acknowledged by Science Research Associates (SRA) for hardening one of the first public networks' security.

Before the 1990s, access to network systems was limited to research groups. As a result, larger vulnerabilities were discovered after networks were made accessible to a wider public through ARPANET (now known as the Internet). In 1988, the government agencies responsible for ARPANET created the CERT (Computer Emergency Response Team). But as the size of the user base of the Internet grew, it was no longer possible to provide custom measures and a team. The measures were shifting toward building antivirus methods and firewalls.

The first generation of firewalls (see Chapter 5) was built in 1987 by engineers from Digital Equipment Corporation (DEC), who later went on to build the first working model at AT&T Bell Labs. These firewalls were focused on packet filters. The second generation of firewalls was built in 1990 with the extra feature of maintaining the endpoints and remembering which port number was being used by the two IP addresses involved in the conversation. The third generation of firewalls came in 1993 and allowed checks for disallowed protocols on an allowed port. From 2012 and onwards, firewalls evolved by providing inspection of application layers using

DOI: 10.1201/9781003038429-11

functionalities like IPS (Intrusion Prevention System), user identity management integration, and web application firewalls (see Chapter 5).

When the first original 802.11 Wireless Network (Wi-Fi) standard was released in 1997, it had Wired Equivalent Privacy (WEP). WEP was supposed to make wireless Internet as confidential and secure as the wired Internet. It needed a 26- or 10-digit hexadecimal key, which was pre-shared. It was not compulsory to use, so unsurprisingly, many people did not use it. When the key was later hashed from human-readable passwords, its usage grew, but it used an easily penetrable weak Rivest Cipher 4 (RC4), also known as ARC4 or ARCFOUR, which meant that the alleged RC4 was a stream cipher.

In 2003, Wi-Fi Protected Access (WPA), which used Temporal Key Integrity Protocol (TKIP), used a new ephemeral key for each packet sent, but was later discovered to be vulnerable to Man-In-The-Middle attacks. In 2004, WPA2 replaced TKIP with AES-CCMP – Advanced Encryption Standard (AES) and Counter Mode Cipher Block Chaining Message Authentication Code Protocol (CCPM). It was later found to be penetrable with a replay attack known as a Key Reinstallation Attack (KRACK). In 2018, WPA3 was introduced and prevented replay attacks by replacing the Pre-Sharing of Keys (PSK) with Simultaneous Authentication of Equals (SAE), which securely identified peer devices with one another.

As of today, the most commonly deployed wireless network security protocols are Wireless Intrusion Prevention Systems (WIPS) or Wireless Intrusion Detection Systems (WIDS). The goal of WIPS is to prevent any unauthorized access of wireless devices to a Local Area Network (LAN) and information assets. WIDS continuously monitors the radio spectrum for any unauthorized access points and activity of wireless attack tools and alerts a systems administrator whenever such an incident is detected.

11.3 HOW NETWORK SECURITY TECHNOLOGIES WORK

ANALOGY

Imagine a system that monitors your activities. This system, when you are online, obtains the location you are connected to, the network you are on, the sites that you visit, how long you stay on each site, what you click and more, to make sure that you are not infected or connecting to a suspicious network and clicking on a dangerous package (Figure 11.1).

Imagine you are being followed by an agent that monitors all your activities: where you go, what you do, to whom you talk, etc. Then this agent will warn you if your action might harm you.

11.3.1 FIREWALL

A firewall is a network security system that observes the flow of incoming and outgoing network traffic (see Chapter 5). It serves as a "wall" between the trusted internal and untrusted external Internet, allowing only the traffic validated by a

FIGURE 11.1 Network security.

predetermined set of rules to pass by. Firewalls make use of the information in TCP and UDP packets for filtering (see Chapter 2).

11.3.2 Intrusion Detection System (IDS) and Intrusion Prevention

An IDS is a software and/or a device that closely monitors a network for any malicious activity. An administrator is alerted when the security information and event management (SIEM) system classifies an activity as an intrusion.

An intrusion can occur at the network or at the host. For a network intrusion detection system (NIDS), the traffic of an entire subnet is monitored. It may use a signature-based detection method, which looks for certain patterns of known malicious instruction sequences or byte sequences. An IDS may use anomaly-based detection as well. This targets malware sequences that have never been seen before. It relies on machine learning to create a model of trusted activity and compares it to the current activity.

For detecting intrusions at the host, a host intrusion detection system (HIDS) is used. It is run on the individual hosts of a network and therefore only monitors the inflow and outflow of traffic at the host. It takes a snapshot of the current file system and compares it to the previously taken snapshot. If it finds that critical files are deleted, added, or modified, then the administrator is alerted.

Intrusion prevention systems function a lot like IDS, as they identify a threat, and as a part of the next step, the network can prepare to protect itself, causing the attack to fail. This can be done in various ways: by going offline, changing the security environment, monitoring the attack content, blocking incoming traffic, etc.

11.3.3 WI-FI PROTECTED ACCESS 3 (WPA3)

WPA3 is the latest wireless network security certificate protocol released by the Wi-Fi alliance. The focus of WPA3 is at the "entry," when a new device requests to connect into the network. It uses the Simultaneous Authentication of Equals (SAE) to oversee this authentication, and the SAE blocks the potential KRACK (Key Reinstallation Attack). It also offers "forward secrecy," that is, even if the private key of the server becomes compromised, the session key will not be. So if the long-term password or secret keys to the network were hacked, the past encrypted sessions cannot be decrypted.

11.4 NETWORK SECURITY TECHNOLOGIES

Network security can be classified into the following three types:

- Physical Network Security: Includes controlled access such as fingerprint authentication, devices, controlled access, routers, etc. (see Chapters 3 and 4).
- Technical Network Security: Controls protected data that is kept on a network and is, or will be, transmitted across or into the network. The data has to have two layers of protection, internal and external. Internal to prevent employees from accessing the data, and external so that the network can defend itself against attackers.
- Administrative Network Security: Includes policies that primarily process and control user behavior. It includes user authentication, role management, access control, infrastructure changes, etc.

The task of implementing network security is in the hands of network administrators, whose primary task is to adopt preventive measures to ensure that the networks they are in charge of protecting are free from any kind of malicious activities and hazards.

Network security becomes even more important with the usage of cloud computing such as Rising IaaS (Infrastructure-as-a-Service), PaaS (Platform-as-a-Service), and SaaS (Software-as-a-Service). These platforms must follow all network security compliances because thousands of SMBs trust them to safely host their personal data.

11.4.1 FIREWALLS

Firewalls are used by individuals and enterprises to protect their servers and user data from incoming network traffic. Firewalls have rules set for allowing or blocking traffic (see Chapter 5).

11.4.2 EMAIL SECURITY

Email is the most common way to spread malicious code and get personal data from devices. Since emails allow scripts to be embedded inside them, it's fairly easy for anyone to write a code, insert it into an email, and send it. This results in the receiver clicking the email and possibly compromising their device. The most effective security methods are email spam filters, which use deep learning to filter domains that send spam or identify email text from millions of emails.

11.4.3 ANTIVIRUS AND ANTIMALWARE SOFTWARE

If a device is on the verge of being compromised, the native antivirus software can help the device stay safe. Since the antivirus software runs constantly in the background of the client, the software can constantly monitor the different types of cybersecurity threats that can occur (see Chapter 9).

11.4.4 VIRTUAL PRIVATE NETWORK (VPN)

A VPN connects any remote device to the official enterprise server where any company's data is hosted. A VPN is the most used software for remote workers, which uses IPSec and SSL to encrypt and authenticate the device. A VPN piggybacks on a public network to directly access the server, such that the public network has no knowledge of the user's activity from the VPN. This is one of the most powerful network security tools.

11.4.5 INTRUSION PREVENTION SYSTEMS (IDS)

IDS have become more important with the rise of SaaS companies, scanning incoming network traffic in real time to proactively detect and block network attacks. If one of the devices is compromised, the IDSs can use mesh networking to easily detect and block attacks.

11.5   ADVANTAGES AND DISADVANTAGES OF NETWORK SECURITY

11.5.1   ADVANTAGES OF NETWORK SECURITY

Network security essentially translates into creating a secure network. The network can be public or private. Network protection requires the prevention of misuse or unauthorized access to the network or its resources. Generally, network security technologies provide various benefits, as listed below:

- Data Protection: A network contains confidential data, such as personally identifiable data, financial information, username, and passwords. Anyone who joins the network will interact with such sensitive data. Therefore, network protection should be in place to protect against malicious users.
- Cyber-Attack Prevention: Almost all cyber-attack attempts on a network come through the Internet. Some hackers are more professionals than others, and the attacks range from small malware to a well-organized distributed denial of service attack (DDoS). These attacks may lead to data breaches and can cost a lot to the company/organization. Protecting the network will prevent these attacks from damaging the computers, the company, and its customers' data.
- Access Control: Access to a network is very critical, and the level of access to a user is important as well. A good network should have multiple access levels to protect sensitive data and resources. The type of access should also be controlled. Network security technologies provide the capability to create granular permission policies for users.
- Centralized Control: Contrary to a desktop security program, a single administrator called a network administrator manages the network security software. This provides centralized control over the entire network. Centralized control provides the capability of changing network security policies with ease.
- Blast Radius: A good network should always have the best network security technologies in use. But it should also be capable of working sufficiently in the case of a breach. Network security technologies can be used to create networks that have minimal blast radiuses, so that even if the system is breached, the impact is minimal, and the attacker is not able to gain control over the entire system.

11.5.2   DISADVANTAGES OF NETWORK SECURITY

While network security technologies provide exceptional advantages and can be crucial for security, there are some disadvantages as well:

- Cost: The process of implementing network security technologies can be costly. In particular, for smaller networks, the purchasing and deployment of security tools can be particularly challenging in terms of cost.

- Time and Resources: The process of securing a network can consume a considerable amount of time. The multi-stage process requires development, testing, deployment, and constant monitoring.
- Human Error: Handling security for large networks is not an easy task. It requires hiring highly trained technicians who can deal with a variety of safety problems and make the network operate efficiently. The administrators must be professionally qualified to fulfill these requirements and because the network administrator has so much access and influence, a small human mistake can turn into a catastrophe for the company.

11.6   WHICH PRODUCTS USE NETWORK SECURITY

Networking security technologies have evolved significantly in recent years. More and more sectors are adopting robust network security to secure IT infrastructure. From a complex data center to the Wi-Fi of a local cafe, every network needs some amount of network security technologies to protect against malicious activities. Below are a few cases:

11.6.1   BANKS/FINANCIAL INSTITUTIONS

- Modern-day banks have very complex IT infrastructure. From big data centers to remote ATMs, every component is critical and may contain valuable information.
- Cutting edge network security tools are applied in the infrastructure to secure it against any probable cybersecurity issue. The banking sector is also a high-risk and high-value target.
- It receives a huge number of cybersecurity attacks each day. Banks need robust firewalls to block these attacks. These advanced firewalls block malicious traffic with the help of unique policies defined by the bank.

11.6.2   HOSPITALS/MEDICAL INSTITUTIONS

- In today's generation, data is one of the most valuable resources in the world. Hospitals contain PID (Personally Identifiable Data) – a subset of personally identifiable information (PII), which identifies a unique individual and can permit another person to "assume" that individual's identity without their knowledge or consent. PID and PII are very sensitive data.
- With countries coming up with strict data regulations, the storage of PID is a challenging task. Hospitals must store their data in the country in which they are located. And recent data breaches have encouraged hospitals to invest in cutting edge network security technologies to protect the data and keep the system operating in any condition.
- In general, hospitals follow various network security protocols and use differing technologies to compartmentalize and store user data. Many hospitals have a centralized software system to operate and manage their medical equipment.

11.6.3   Cloud Computing

- Cloud computing has grown exponentially since its introduction. From a small-scale start-up to a large enterprise, every business is adopting cloud technology to improve scalability and to reduce maintenance and cost.
- Cloud computing leverages virtualization technology to use hardware more efficiently by sharing the hardware between multiple tenants. Virtualization makes use of a hypervisor and multiple VM (Virtual Machines) to share physical resources.
- The resource sharing improves efficiency, but it creates security concerns. Along with the existing network and security challenges of a typical system, virtualization has additional challenges like Virtual Machines (VM) to VM and VM to hypervisor injection.
- Network security plays a huge role in preventing these kinds of attacks. Hypervisors need to be equipped with extra network security tools compared to a typical operating system.

Apart from internal network security, cloud computing also faces a big challenge of external network security in terms of data center breaches. A cloud computing data center handles a huge amount of traffic from various parts of the world. It is also a high-value target as it hosts multiple tenants. These data centers are equipped with multiple firewalls, access control technologies, and other network security tools to protect against possible attacks.

While network security technology can be used in a wide variety of applications, the technology has some constraints also. According to the Open Systems Interconnection (OSI) model, network security technologies can't be applied beneath the network layer. As a result, the physical layer and data link layers aren't equipped with network security technologies.

Network security technology cannot work on physical devices, which means that if a machine is accessed physically by a device like USB, network security technology will not work, and the system may be at risk. If a malicious person gains access to the machine physically, then he/she would be able to bypass the network security tools and connect to the machine and the network and thus have access to the data available in the system.

11.6.4   Use of Network Security Technologies

The major areas where Network Security technologies are used are as given below.

11.6.5   Firewalls

A firewall is a network security technology that protects a private network from unwanted, unauthenticated, and unauthorized traffic at both entry and exit level (see Chapter 5).

Cisco and Palo Alto Networks (PAN) are the major companies providing both enterprise-level and home firewalls. Cisco ASA (Adaptive Security Appliance) provides the combined features of firewalls, antivirus and intrusion prevention, and provides customers high performance due to multi-site and multi-node clustering techniques. Palo Alto Networks NG firewalls give in-depth granular control over the network. The App-ID feature in PAN firewalls restricts applications based not only on protocol and port number but also on the application's behavior and the user interaction.

11.6.6 ANTIVIRUS

Antivirus programs act as a gatekeeper to an endpoint, protecting it from all possible threats and attack vectors. Sophos Endpoint Protection and Crowdstrike are some of the leading companies providing next-generation antivirus solutions. The advanced anti-ransomware feature and complete support to MacOS in Sophos makes it unique among its competitors. Crowdstrike protects the endpoints by employing advanced machine learning and artificial intelligence signatureless technologies to provide network security from an endpoint context. Antivirus solutions protect endpoints from virus, unwanted spams and ads, and warn the user (see Chapter 6).

11.6.7 SECURITY INFORMATION AND EVENT MANAGEMENT

SIEM is software that collects and aggregates log data generated throughout an organization's technology infrastructure, from host systems and applications to network and security devices such as firewalls and antivirus filters. The software then identifies and categorizes incidents, analyzes them, and aggregates all events generated in the network – from hosts, network security appliances, and applications. Splunk and Exabeam are currently popular SIEM products in the industry. Splunk SIEM is built based on an advanced Big Data platform, thereby helping in improving an organization's overall cybersecurity strength, detection and investigation and providing in-depth analytics. A data lake, advanced analytics, and Security Orchestration and Automated response components in Exabeam SIEM assists analysts and investigators in deep diving into security incidents. SIEM supports hot and cold log storage. SIEM can cease malicious activities that are not captured or missed through traditional defenses.

11.7 CONCLUSION

This chapter highlighted the different types of network security and their workings in order to emphasize the importance of network security. Network security reduces the attack surface (the sum of the different points of attack where an attacker can try to enter data or extract data from an environment) of an organization and acts as a defense from intruders and malicious activities. Defense in Depth, an approach of using multiple security technologies at various layers in an organization, provides complete visibility to an administrator and protection from different types of threats. A single network security technology alone is not sufficient to overcome the current attacks taking place in the world. Traditional network security technologies were signature-based whereas current network security technologies are behaviorally based and are designed mostly using advanced machine learning models. Maintaining the CIA triad (confidentiality, integrity, and availability) and protecting data in rest and data in transit are the key guidelines for securing a network. Thus, we can conclude that an effective network defense is essential for business continuity. Figure 11.2 shows the taxonomy of network security.

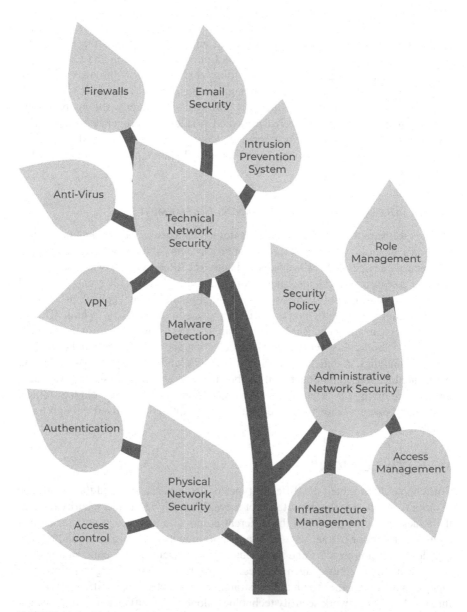

FIGURE 11.2   Taxonomy of network security.

**SAN JOSE PROGRAMMER PLEADS GUILTY
TO DAMAGING CISCO'S NETWORK**

In August 2020 a programmer pleaded guilty to hacking into Cisco's network and hurting the company's operations.

The programmer, a former Cisco employee who resigned from his position, admitted that he logged onto Cisco's servers and deployed a code from his Google Cloud Project account and caused the deletion of 456 virtual machines for Cisco's WebEx Teams application. The programmer's code shut down 16,000 WebEx Teams accounts for two weeks [3].

12 Location Tracking

12.1 INTRODUCTION

The emergence of the Internet connecting the entire globe at a scale and magnitude never before seen among the human population. As a result, we have access to all information through a simple click. As technology continues to improve rapidly, the machines and devices that we use have become so smart that they don't require even the minimal work of clicking to give us the information that we need. Our devices are now well equipped with hardware and software components that sense our movements, track our paths, analyze our routes, and go the extra mile to suggest what to do at a certain place. One of the most important technologies behind this is location tracking.

Location tracking technologies locate, record, and track the physical movement of people or objects and are ubiquitous and in use by many of us every day. The most commonly used location tracking service is the Global Positioning System (GPS). Among its many uses include in car navigation systems, location identifiers on digital pictures and for businesses close to our respective locations such as restaurants, shops, service providers, etc., through the help of popular mobile applications like Yelp, Zomato, Uber, etc. We always assume that location tracking is only done through our smartphones and the GPS in them. However, it is important to understand that there are many other ways through which locations are being tracked and used.

Bluetooth Low Energy (BLE) or Radio Frequency Identification (RFID) tags, Quick Response code (QR) and Wi-Fi networks are all location trackers. BLE and RFID work like a beacon that can share our location information to a controller. QR codes, a basic alternative to a barcode, can be a tracker if it stores the details of the location where the QR code is scanned. A common example is a misplaced product on the shelves of a huge store or a product being shipped to the wrong facility. As we use a variety of Wi-Fi networks offered at public buildings, malls, restaurants, coffee shops, libraries, and more, the networks track our location and keep a record of it.

After a brief historical background, this chapter will review GPS navigation system using satellites that is used for location tracking, explain how the technology works, the advantages and disadvantages of the different location tracking technologies, and the different products which use this technology.

12.2 BRIEF HISTORICAL BACKGROUND

In 1973, The US Department of Defense started to work on the Research and Development of location tracking systems and decided to incorporate or use satellites. The five-year-long research project culminated with a working model in 1978. The prototype worked well, and the United States started to add a greater number of satellites into orbit, with a total of 24 satellites by 1993. These satellites enabled

DOI: 10.1201/9781003038429-12

the US government to build surveillance systems that successfully covered the entire globe. While the use of GPS was initially restricted only for military purposes, in 1983 President Ronald Reagan signed an executive order that allowed GPS to come into use for the public. From then on, people worldwide started to explore and use GPS systems, and they have now become the basis for many products, business models, and applications.

The first car navigation systems were created by Alpine and Stanley Electric (now Honda) in 1981. In 1989, a new system was developed that could process five important functions: the location of the vehicle, storing routes and destinations in the database, instruction generation, planning a better route, and vehicle control [1].

12.3   HOW LOCATION TRACKING TECHNOLOGIES WORK

ANALOGY

Imagine an invisible person who follows you all the time and records your location by capturing the latitude and longitude of your position on earth. This invisible person not only monitors your location, but also records and memorizes what you usually do in that location, whether it be exercising, working, sleeping, attending class, etc. In this case, the invisible person is your mobile phone tracking your geolocation (Figure 12.1).

The 24 satellites orbiting in space cover the entire earth and they are collectively called NAVSTAR. At any place in the world, the user of GPS will be able to access a minimum of three and a maximum of four satellites. Since there are so many satellites covering all of the earth's surface, GPS is very accurate (down to within a foot of the location) and works precisely with much more complex devices or equipment.

GPS mainly works by using a method called "triangulation" or "trilateration." The satellites in space are kept as reference locations to identify the latitude and longitude of any point on earth. Any GPS device must connect to three or four satellites at a time for proper functioning. To "triangulate," a GPS receiver first measures the distance between itself and each satellite. It uses the simple formula of distance = time of arrival * the speed of light to measure the distance. The distance is calculated by knowing how much time the GPS receiver takes to receive the signal from each of the connected satellites. The travel time is calculated at the receiver end with the help of atomic clocks that are installed in the satellites. Along with the distance and time, the receiver constantly obtains information on the location of the satellite. So, by combining all three factors, the GPS receiver makes a calculation and displays the exact location [2].

12.3.1   USES OF GPS TECHNOLOGY

There is an enormous number of possible uses for GPS. These systems are extremely versatile and can be incorporated in almost any field known. They can be used to

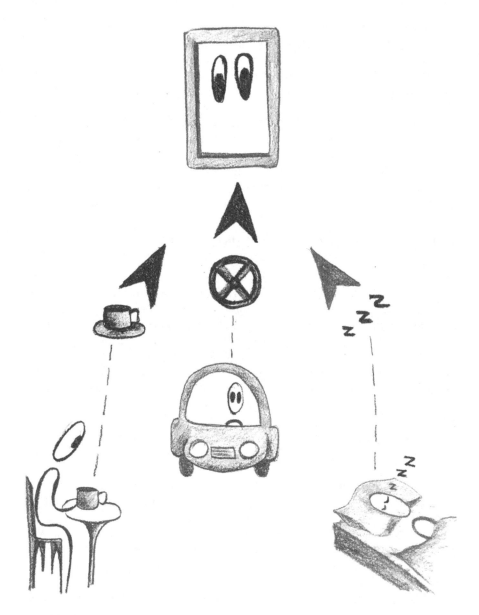

FIGURE 12.1 Location tracking.

locate different kinds of landscapes, help in the field of agriculture and navigate cars on the ground, as well as airplanes in the air. GPS has become the most reliable military application technology, helping to locate and identify targets and people in need during emergencies.

There are five main categories for the applications of GPS, and they are calibrated accurately every day. These include:

1. Identifying the location of a person, target, or thing
2. Navigating from a source to a destination – used in transportation
3. The process of observing and monitoring a movement, called tracking
4. Storing information to create maps
5. Precisely measuring time

There are four Global Navigation Satellite Systems (GNSS). GPS (owned by the United States), GLONASS (Russia), Galileo (EU), and BeiDou (China). Additionally, there are two regional systems – QZSS (Japan) and IRNSS or NavIC (India).

In 2020, China sent its own satellite to space, which allowed it to complete a global navigation network and wean it off of US technology. This development ensured China would not have to depend on the US satellite system [3].

12.4 LOCATION TRACKING TECHNOLOGIES

12.4.1 GPS TECHNOLOGY

The GPS of the United States consists of up to 32 medium Earth orbit satellites in six different orbital planes, with the exact number of satellites varying as older satellites are retired and replaced. Operational since 1978 and globally available since 1994, GPS is currently the world's most utilized satellite navigation system.

GPS allows users to find their exact location, speed of their travel and time, and can be widely used for navigating on land, in the sea or by air. It is very robust and can be used in all types of weather conditions and in any corner of the earth. GPS is known not only to support the public but is also highly used in the military and in commercial products and applications.

GPS mainly uses the satellites that revolve around the earth. There are 24 GPS satellites, among which 21 are active, and three are kept as a backup resource. They are located approximately 11,000 miles above the earth's surface and are spaced in a way that four satellites will be above the horizon. Each of these satellites is equipped with high computing devices, an atomic clock, and a radio. These satellites constantly update their position and time settings every day to the ground station. The satellite system makes GPS more accurate and free from minor errors. With the help of these satellites revolving around the earth, we can identify the latitude and longitude information about the user, and these details will be forwarded to the person via software or an application.

The user is a receiver and can obtain a visual map of their location if they have a display screen. If the user is moving, the information on their speed, direction of travel, and the time to reach a particular destination can be known. The data collected from GPS systems is extensively used for creating maps in Geographic Information Systems (GIS).

12.4.2 RADIO FREQUENCY IDENTIFICATION

RFID is one of the location tracking technologies widely used in supply chain, health care, and offices to track people moving within confined environments. RFID uses small microchips that are physically affixed to the objects whose movements are to be tracked. RFID tags are a passive tracking device and only transmit the location of the item when activated by the reader. The reader sends RF waves that prompt the RFID tag to communicate with the reader on a predetermined RF frequency. The data is then sent to a centrally located database.

12.4.3 INTERNET TRACKING

Internet tracking is usually used to track and recover lost or stolen laptops, mobile phones, and other computing devices. Nowadays, most computing devices come with an optional location tracking system which, if activated, connect to a central server whenever there is an Internet connection. This method provides valuable information like the IP address and wireless data for geographic triangulation and finds the devices' exact location for recovery.

12.4.4 ADVANTAGES AND DISADVANTAGES OF LOCATION TRACKING

Even though location tracking can make life more convenient and comfortable by helping us find houses, seek out restaurants, book transportation, and all the myriad other ways it is used in everyday life, there is a growing feeling among people of the violation of privacy. The fact that large corporations like Facebook or Google might know all of our movements make people more conscious and worried about their private space. Therefore, it is crucial to analyze how important it is to adapt and continue to use these technologies. The positives and negatives of using location tracking systems must be weighed against the comforts, needs, and privacy of people.

12.4.4.1 Advantages of Location Tracking

12.4.4.1.1 Emergency and Safety

GPS tracking systems are the most accurate and fast in identifying locations. GPS on mobile phones can help to track people who require help in an emergency or also to assist the elderly in a smart home environment [4]. It can also be used to track children or vulnerable adults traveling alone. Some applications allow users to share their locations with selected people over a particular period. Restricting location sharing can increase safety and security, assuming measures are taken to protect privacy and prevent location information from falling into the wrong hands.

12.4.4.1.2 Tracking Phones

Smartphones are increasingly in the hands of many people throughout the world and serve a number of important functions. Yet, because they are very expensive and store large quantities of valuable and private information, they are that much more important to track when they are lost or stolen. GPS receivers are now capable of tracking the location of our phones and providing real-time updates on their status and location. Phone tracking can help people locate their lost or stolen devices and can also be very useful for security forces to track people in the case of an emergency (as explained in the previous section).

12.4.4.1.3 GPS in Cars

Cars use location tracking systems like GPS to navigate and locate cars remotely (e.g., searching for a car in a crowded and multiple level parking garage). These tracking systems have become more sophisticated and can call for help during emergencies or allow people to unlock cars remotely, if the physical key is missing.

12.4.4.1.4 Networking

Location tracking can provide many benefits to people due to the numerous ways in which they use social media. For example, location tracking in dating apps can increase the sense of safety for someone going on a date with a stranger. A similar example could be ridesharing, where there is an assumed sense of security for the passenger because they know that their location is being tracked, thus making it harder for a driver to commit a crime or take them to an undesired location. A final example could be for travelers to use GPS and tracking to inform their loved ones of their location, so as to ensure them of their safety. There are many more potential examples to note but these are a few.

12.4.4.1.5 Memorable Events

When people take pictures, their location is automatically saved. This feature helps people to know the location of photos, even years after they were taken. This extends to events and occasions being recorded by time and location, which allows people to easily track and create memorable events over time (e.g., photo album for children, memories from a vacation, etc.).

12.4.4.2 Disadvantages of Location Tracking

12.4.4.2.1 Stalkers

One of the main downsides of location tracking is attracting unwanted attention from people or, in more extreme cases, stalkers. Although it can be helpful to know the whereabouts of a family member, using location tracking can allow someone to become victims of stalking. This is mainly due to the improper use of the permissions function within applications or their feature that allows anyone and everyone to access location information.

12.4.4.2.2 Advertisements

Location tracking and browsing history are increasingly used to create targeted advertisements that we see every day in our applications, websites, and other Internet use. Companies store and use our location information in their databases that they can use for their advertising campaigns. More advanced machine learning and data analytics algorithms build a portfolio for users and are sold to advertising companies [5].

12.4.4.2.3 Theft

As our devices keep tracking us, any kind of public or leaked check-in information attracts thieves who have detailed knowledge of our schedules. There are many known cases of thieves and criminals targeting an individual based on their travel or location information.

12.4.4.2.4   People Movement Control

Location tracking apps also might be used by government and authoritarian regimes to control the movement of a specific group of people. A clear example of this is the "Absher" app that is used in Saudi Arabia by male users to log the location information of their dependents and thereby grant or revoke permissions for them to travel or perform certain tasks with the click of a button [6].

12.5   WHICH PRODUCTS USE LOCATION TRACKING

Numerous apps and products use GPS location tracking services and can be categorized based on their applications.

Maps are the most commonly used products of GPS. For example, Google Maps uses our real-time location to provide routes for destinations.

Similar to maps, the transportation sector also uses location-based tracking GPS to its advantage. Uber and Lyft are two giants dominating the renewed taxi business, which solely depend on location tracking systems. In addition to this, carpooling apps like Scoop that automatically match commuters in the same area for carpooling use location tracking to identify people commuting to the same place regularly. The recent example of on-demand and short-term electric scooters and bike rentals (increasingly common in cities around the world) also rely fully on location tracking. Examples include Ford bikes and Bird or Lime electric scooters. These companies rely on GPS to allow users to locate the docking stations (where the bikes or scooters are located) and can even provide real-time bike status.

In addition to the above products, GPS is also used in Apple iPhone, Google Pixel, and many other smartphones to automatically tag photos with their location. This geotagging is used by many networking apps like Instagram or Facebook to update the user's homepage as they upload pictures.

Since there are numerous advantages of using GPS, there are many products in the market that are solely dependent on it to provide users with greater convenience.

These applications can be categorized accordingly:

- Location trackers: Google Maps, Waze, Glympse
- Social networks and messengers: Facebook, WhatsApp
- Dating apps: Tinder, Hinge, OkCupid
- Fitness apps with GPS route tracking: Nike Running Club, MyFitnessPal
- On-demand delivery apps: Uber, UberEats, Postmates, GrubHub
- Travel apps: Expedia, TripAdvisor, Hotels.com
- Weather apps: Yahoo Weather, Weather Underground
- Photo location finder apps: Google Photos, Explorest
- Meet and Greet apps: Meetup, Nextdoor, Bumble, Peanut, Skout, Nearif, Meet My Dog, Foursquare City Guide

The abovementioned apps are only a few of the thousands of applications in the market that use GPS.

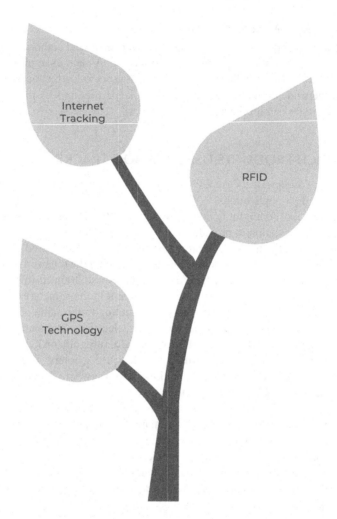

FIGURE 12.2   Taxonomy of location tracking technologies.

12.6   CONCLUSION

Over the past few decades, technologies like GPS, RFID, and Internet tracking have improved dramatically. Consumers and businesses alike have benefited from these tracking technologies. Like any other technology, tracking devices face challenges when it comes to the security and exploitation of private data. Furthermore, location tracking is not just limited to people but could also be used to track wild animals as well. For example, researchers or conservation agencies can remotely observe wild animals for the purposes of protection or research.

Moving forward, it is essential to better understand who has the right to access this location data and what actions need to be taken when this data gets compromised. Tracking technologies are not limited to GPS or RFID but can also open doors to many other advances, if used wisely (Figure 12.2).

WIFE-TRACKING APPS

"Absher" is a Mobile Application in the Kingdom of Saudi Arabia which allows men to access the travel abilities of their spouses or other women or workers they employ. With minimal effort, men can allow or withdraw permission to travel abroad, cancel airline tickets, or register for SMS updates if an unauthorized escape is attempted. "It gives men control over women," according to the Guardian report [7].

CORONAVIRUS PANDEMIC AND PEOPLE TRACKING

When the Coronavirus pandemic started in early 2020, the importance of tracking people affected by the virus and quarantine became a life-saving strategy. Several initiatives began in the United States and many countries to track people's location.

According to the *Wall Street Journal*, government officials across the United States started using location data from millions of cellphones to better understand the movements of people during the pandemic. Interestingly the data for the movement of people did not always come from cell phones but often came from the mobile advertising industry. The data helped to show which retail establishments, parks, and other public spaces were drawing crowds that could risk accelerating the transmission of the virus [8].

13 Surveillance

13.1 INTRODUCTION

Surveillance technology is an important area in cybersecurity and one of the most representative products of the new century's technological progress. Although the formal definition of surveillance – the monitoring of behavior, information and activities – is abstract, the applications of surveillance are ubiquitous. From security cameras to cybersecurity software, surveillance technologies have been deployed everywhere for a variety of reasons.

Although the technology has dated as far back as World War II, it has evolved throughout human history and played a major role in monitoring and data analytics of data. For example, in World War II, surveillance technology took the form of spy glasses, recording devices, and radios for spying and espionage. Later in the 20th century, video surveillance such as CCTV was used for public and home security. And in the modern century, surveillance has branched into cybersecurity to identify security threats, vulnerability, mitigate risks and to shield against hacking and malware. In today's world, surveillance is a significant aspect of society, so much so that author Shoshana Zuboff refers to our current time as, "The Age of Surveillance Capitalism" [1].

After a brief historical background, this chapter will review how surveillance technology works, discuss the most recent applications of cybersecurity with surveillance, the advantages and disadvantages of surveillance technologies, and its use among products.

13.2 BRIEF HISTORICAL BACKGROUND

Although surveillance technology's widespread use seems to suggest that it is a product of the 21st century, surveillance has taken different forms throughout human history. One of its earliest forms was wiretapping and began long before the telephone. As early as 1862, California published a statute prohibiting wiretapping. This came as a response to the conviction of D.C. Williams, a stockbroker who was convicted of eavesdropping on corporate telephone lines and selling information to stock traders when the Pacific Telegraph Company had established its lines on the West Coast [2].

In the 1930s, wiretapping was mainly used by private detectives for investigation and for corporations as a tool to root out union activities. According to the La Follette Civil Liberties Committee of the United States Senate (1936–1941), wiretaps were heavily used to spy on union laborers [3]. Aside from professionals and large corporations, the government also used surveillance and wiretapping.

The most infamous example could have been Watergate, where the Republican executive branch was using wiretaps on the opposing Demcratic party to gain

DOI: 10.1201/9781003038429-13

political benefits. This resulted in the impeachment and resignation of President Nixon and also spiked public concern about wiretapping.

Aside from wiretapping, surveillance was also noted and referred to as espionage or spying. It was used as a means to gather and collect information and observe the actions of enemies. These developments go back to early examples including spy glasses and telescopes in the 1600s as well as radio surveillance in the 1890s. One of the most noticeable technologies was video surveillance, which dates back to World War II.

In 1942, Walter Burch, a German engineer, invented Closed-Circuit Television (CCTV). Soon thereafter, it was used by the German military to observe rocket launches inside bunkers. The US government also used CCTV to observe and test nuclear bombs without having to come into contact with the dangerous aftermath. A decade or so later, CCTV had been adopted into households for security purposes. For instance, the British installed CCTV technology to ensure security when the Thai royal family visited Britain in the 1960s [4]. Later, the London Transport system started to deploy CCTVs throughout train stations to increase public safety. In 1969, Marie Van Brittan Brown received a patent on her system of peepholes and cameras that could allow someone to access and monitor each individually. The cameras would broadcast their images to a monitor. The public had taken an interest in the technology, which resulted in CCTV branching into home security. Banks and retailers began to use CCTV as an additional security measure against theft, and this trend would continue through the 1980s [5].

In the current century, the use of video cameras and monitoring proliferated and spread across countries, from family-owned businesses to restaurants to office spaces and homes. The technology had advanced so much that new technologies emerged out of surveillance. One such example is facial recognition. This technology identified human faces and significantly improved video surveillance. Business owners could identify and register names with faces and be alerted when a particular face came into view. It also became useful for police departments, who could keep a forensic database for law enforcement and criminal investigations [6].

13.3   HOW SURVEILLANCE TECHNOLOGY WORKS

ANALOGY

These days, it seems that our lives more closely resemble those of the characters in George Orwell's novel *1984*. Instead of being monitored by "Big Brother" like the characters in the book, today we are monitored by countless programs, apps, and companies, capable of tracking where we go on the Internet, what movies and shows what we watch, where we travel to, what news sites we read, and more (Figure 13.1).

Surveillance technology can be useful for different parties in different ways. For example, governments can use it for collecting intelligence, counterterrorism,

FIGURE 13.1 Surveillance.

preventing, or investigating crimes, protecting people and more. On the other hand, this same technology can also be used for conducting crimes, collecting intelligence from competitors, international espionage or detecting heresy and heterodoxy within religious organizations. With regard to the development of the Internet of Things (IoT), surveillance technology can also be used to gain access to buildings or networks, monitoring or tracking people, etc.

13.3.1 VIDEO SURVEILLANCE AND CRIME

The primary, but by no means the only, use for video surveillance is for security and safety. This comes in the form of surveillance increasingly being used to monitor, capture, and prevent crime. These days, cameras are placed in public places, stores, and homes to continuously capture any and all activities, with the main objective being to reduce crime.

That said, the evidence so far has not shown that video surveillance does indeed reduce crime. One example is that suicide attackers won't be afraid of video cameras but might be attracted to the contrary. Also, according to a study in Britain, there is no evidence that cameras would reduce the crime rate. However, the video surveillance system is very expensive, which is almost 20% of Britain's criminal justice budget. What's worse, the video record is too boring to watch, and most people cannot pay attention to it for more than a few minutes.

Another reason is that video surveillance invades people's privacy; people are very uncomfortable with being observed without consent. Besides, there is no efficient control over the usage of video surveillance technology. In this case, this surveillance can be easily abused, which might disrupt a citizen's privacy as a consequence. Certainly, we need more refined laws to limit the abuse of surveillance technology and protect citizens' privacy.

Also, video surveillance can't be fully trustworthy for identifying people. For example, traffic cameras cannot be perfectly reliable in identifying drivers, especially while drivers wearing sunglasses, caps, or other disguises. So, people might still have chances to get the wrong traffic tickets.

13.4 SURVEILLANCE TECHNOLOGIES

Surveillance technology is the monitoring of computer activity and data transfers over networks. Such monitoring might be conducted by the government, organizations, or individuals and can be legal or illegal. Using advanced software and biometrics technology, it allows a party to detect threats, catch potential hazards, and identify criminal activities [7].

The Internet is the most common area of surveillance. Network requests and traffic are monitored in real time. Data packets transferring between computers can easily be captured by an Internet Service Provider. By collecting these chunks, any sensitive or criminal information can be analyzed, and the involved party can be traced by their IP address.

Corporate surveillance is also very common. Companies tend to collect users' data for marketing purposes and targeted advertising. For example, when a user searches a brand of a certain product on the Internet, this information can be recorded by a third party. And when this user goes to their Facebook or a seemingly unrelated website, they might see the advertisements of the product that they just looked up or similar products. With business intelligence technology, marketing can be done more effectively, which promotes sales and profits.

Surveillance can also occur from unauthorized websites or malicious software. If a computer is infected with a virus or trojan, malware can scan the hard drive for suspicious data and track the usage and activities. Once the malware is installed on a user's computer, then the hacker can monitor when the user is connected to the Internet and also control the camera, microphone, and other computer devices.

Also, there are a lot of unauthorized websites that represent themselves as real ones. If a user inputs their personal information and password into these websites, their account information or credit card information can be stolen, which can lead to identity theft and financial losses.

There are many surveillance methods, including computers, telephones, cameras, social network analysis, biometrics, Aerial, Corporate, Data mining, and profiling. Surveillance technology works differently within the different methods.

13.4.1 COMPUTER

Most computers monitor data and traffic on the Internet to meet legal requirements. Instead of searching through the vast amount of data manually by human investigators, nowadays automated surveillance computers can identify suspicious data from intercepted Internet traffic and report it to human investigators. This interception is usually triggered by certain words, certain types of websites, email, or online chats with suspicious people. Some law enforcement and intelligence agencies, such as the FBI and NSA, spend billions of dollars annually developing, purchasing, implementing and operating systems to intercept, analyze, and extract useful information [8, 9].

Because they store so much personal data, computers have become a significant target of surveillance. One kind of software, for example, the FBI's Magic Lantern and Computer and Internet Protocol Address Verifier (CIPAV), can be installed remotely, via an e-mail attachment or other operating system vulnerabilities, and can provide unauthorized access to data for the hacker. Van Eck phreaking is another form of computer surveillance where devices can read electromagnetic emanations from computers hundreds of meters away, in order to extract data. Pinwale is a database run by the National Security Agency (NSA), which stores large amounts of emails of American citizens and foreigners. Also, PRISM is a data mining system run by the NSA as well. It allows the US government access to information directly from technology companies, such as search history, live chats, emails, etc. [10, 11].

13.4.2 TELEPHONES

The Commission on Accreditation for Law Enforcement Agencies, Inc. (CALEA) purpose is to enhance the ability of law enforcement agencies to conduct lawful interception of communication by requiring that telecommunications carriers and manufacturers of telecommunications equipment modify and design their equipment, facilities, and services to ensure that they have built-in capabilities for targeted surveillance, allowing federal agencies to selectively wiretap any telephone traffic; it has since been extended to cover broadband Internet and VoIP traffic. Some government agencies argue that it covers mass surveillance of communications rather than just tapping specific lines and that not all CALEA-based access requires a warrant.

Similar to computers, telephone surveillance is very common as well. The Communications Assistance for Law Enforcement Act (CALEA) allows federal agencies to selectively wiretap any telephone or VoIP traffic by requiring telecommunications carriers to include capabilities for targeted surveillance [12]. Additionally, the FBI has contracts with AT&T and Verizon Wireless to get phone call records and search histories [13]. In this process, audio-to-text software translates intercepted audio to machine-readable text. Then, call-analysis programs can search for certain words or phrases and decide whether to involve a human agent [14].

In the United States and the United Kingdom, tools that were originally developed by the military for counterterrorism purposes have now become more widespread. An example is StingRay, a program that can listen to conversations if a phone user is close to microphones activated remotely. StingRay uses invasive cell phone surveillance devices to mimic cell phone towers and send out signals to trick cell phones in the area into transmitting their locations and identifying information. Since StingRay broadcasts signals as strong as cell phone towers, it causes nearby cell phones to transmit their International Mobile Subscriber Identity (IMS) number to them without the user's attention. Thus, an individual's location can be easily determined by their mobile phone. Another technology, called multi-alteration, calculates the time from the phone owner to nearby signal towers [15].

13.4.3 AERIAL SURVEILLANCE

With the progress and development of drones, sophisticated imaging and satellite system-based aerial surveillance technology has gradually become of greater importance. Governments, military, spy agencies, cities, and private industries use these technologies for a variety of purposes.

Aerial surveillance systems work by using small Unmanned Aerial Vehicle (UAV) with optical sensors and image transmission modules and a wide range of technologies involving communication, control, sensing, image processing, and networking [16, 17].

13.5 ADVANTAGES AND DISADVANTAGES OF SURVEILLANCE TECHNOLOGIES

13.5.1 ADVANTAGES

The primary, but by no means the only, use for video surveillance is for security and safety. This comes in the form of surveillance increasingly being used to monitor, capture, and prevent crime. These days, cameras are placed in public places, stores, and homes to continuously capture any and all activities, with the main objective being to reduce crime.

One of the advantages of using surveillance technologies is security. It gives a sense of safety, particularly when used for protecting homes or businesses in areas with high crime rates. Most surveillance systems are wireless, which allows the users to monitor the homes/offices remotely on smartphones or tablets. Surveillance systems might help in preventing theft or improve productivity and punctuality – for example to know when employees have arrived to their place of work.

The fact that cameras can be watching people at all times can deter illegal activities or delinquent behavior.

Since surveillance technologies come in different sizes and shapes, it is potentially easy to use them in a number of different ways. Cameras and video surveillance can be placed in obvious locations where they can monitor extensive activity, but they can also be hidden and harder to notice. So, if you want to take preventive measures, then the mountable surveillance system can be used, but if you want to gather information or if you want to use it for spying purposes, then the small hidden cameras can be used.

A very beneficial use of surveillance technology systems is for gathering evidence and clues related to committed crimes. Video footage is a powerful form of evidence when it comes to proving the guilt or innocence of an alleged criminal. This is only furthered by technological advances like night vision, which can capture the footage of a crime even in dark areas.

Surveillance technology also brings convenience for day-to-day life. For example, cameras installed on traffic signals can prevent people from disobeying laws and running through red lights. These systems can also be used by authorities to monitor real-time traffic conditions and make changes or suggestions according to it. All modern surveillance technologies have a lot of useful features, such as detecting motion and sound, built-in sirens, motion-sensing floodlights, and sending alerts to smartphones when witnessing any unusual activity [18].

13.5.2 DISADVANTAGES

The main issue with surveillance systems is privacy. Since the entire idea of surveillance is watching people without them knowing, it occurs without people's consent. Thus, an invasion of privacy is a challenge when it comes to implementing surveillance systems. People are very uncomfortable with being observed without consent. Besides, there is no efficient control over the usage of video surveillance technology. In this case, this surveillance can be easily abused, which might disrupt a citizen's privacy as a consequence.

Surveillance technologies can also be easily abused. Video captured from a surveillance system installed in public places can be used by law enforcement to blackmail an individual. For example, "in 1997, a top-ranking police official in Washington, DC was caught using police databases to gather information on patrons of a gay club. By looking up the license plate numbers of cars parked at the club and researching the backgrounds of the vehicles' owners, he tried to blackmail patrons who were married" [19, 20]. Certainly, we need more refined laws to limit the abuse of surveillance technology and protect citizens' privacy.

Also, video surveillance can't be fully trustworthy for identifying people. For example, traffic cameras cannot be perfectly reliable in identifying drivers, especially while drivers wearing sunglasses, caps, or other disguises. So, people might still have chances to get the wrong traffic tickets.

Another disadvantage is that surveillance systems can be very costly for many people or circumstances. They can cost thousands of dollars depending on the types of cameras, monitoring systems and features they employ. Installation and maintenance can also add cost. And because a surveillance system is only capturing footage and not actually stopping or preventing a crime from occurring in real time, some might prefer to find alternative security measures.

It can also be used for illegal spying activities [21]. There is also significant vulnerability inherent in video surveillance systems. As users continue to update their methods and use better technologies, intruders can stay up to date with these systems and use loopholes to avoid being detected. Since cameras tend to be stationary, if offenders know of the placement of the surveillance system, he/she can easily avoid the cameras altogether. Savvy criminals who know of these technologies can disconnect surveillance systems from their power source using different ways, or in the worst-case scenario, they can use surveillance systems to spy on the owner/surveiller.

Hackers can attack surveillance technologies by using the interference in the signal which can result in recording failure. If some object is present between the camera and the offender, then there is no use in recording it. Using security cameras cannot guarantee the complete prevention of theft. Video surveillance can record footage and potentially identify criminals, but it cannot actually stop it at the time of criminal activity [22].

13.6   WHICH PRODUCTS USE SURVEILLANCE

Although surveillance has been a major concern for the general public and is viewed as a violation of privacy by many civil liberty activists, it has evolved consistently with modern technologies, especially with the emergence of the IoT.

Computer surveillance has been established ever since the invention of the Internet. The majority of computer surveillance involves monitoring and collecting Internet data and traffic. For instance, in the United States, under the Communications Assistance For Law Enforcement Act, all the phone calls and Internet traffic, such as emails, instant messaging, are required to be available for real-time monitoring by federal law enforcement. Billions of dollars are spent by US agencies like NSA and FBI to purchase, implement, and operate automated Internet surveillance technology (like Carnivore) to intercept and analyze data. NSA also runs a data mining system like Pinwale that extracts and stores the emails of American citizens and foreigners.

Aside from computer surveillance, surveillance has been frequently used in telephones and cameras, as discussed in the previous sections. Mobile phones are commonly used to collect location data and can be used to determine whether a phone is being used or not. On the other hand, surveillance cameras are used for monitoring a specific area and are generally connected to a recording device or IP network for security and law enforcement.

Social networking companies like Facebook and Twitter also use surveillance to create social network maps. These maps are then used by data mining technologies to extract useful information such as personal interests, friendships, beliefs, thoughts, and activities for commercial purposes.

Surveillance can also take form in biometrics, a technology that measures and analyzes human characteristics for screening, authentication, and identification. For instance, facial recognition uses a person's facial features to identify one's identity. Both the Department of Homeland Security and DARPA are heavily invested in facial recognition systems. Another form of such surveillance is DNA profiling, which looks into somebody's major DNAs to produce a match.

Aside from that, there is aerial technology, which is the gathering of imagery and video from airborne devices or vehicles. Surveillance has been deployed on un-crewed aerial vehicles, spy planes, and helicopters, where military aircraft use different sensors to monitor the battlefield. Many other modern technologies such as digital imaging and miniaturized computers have contributed to advances in aerial surveillance hardware such as micro-aerial vehicles and forward-looking infrared.

Finally, surveillance is driven by emerging industries like data mining – the application of statistical techniques and algorithms to discover previously unnoticed relationships within the data – and data profiling – the process of assembling information about a particular individual or group to generate a profile. Most financial and transaction records are electronic today. Credit card payments, using a bank ATM or a phonecard, generate electronic records. Public records such as court records, taxes, and other records are digitized and are often made available online. All of this information is surveyed, collected, and stored for data mining and profiling.

13.7   CONCLUSION

The development and propagation of surveillance technologies have had, and will continue to, a significant impact on most people and in nearly all aspects of their private and public life, be it at home, at work or in public places. Surveillance is not just limited to home security cameras but also includes public cameras, drones, and aerial surveillance by public and private companies. The large-scale collection of mass data can be a source of concern for many citizens in any society. Businesses and homeowners are more likely to feel safe in the presence of cameras. They can give them the peace of mind that their property is secure and safe, especially in areas where crime rates are very high [22]. While in some cases (i.e., a theft or burglary), a surveillance system can provide useful information for finding the offenders and tracking down the stolen valuables, there are valid concerns among many citizens that unprotected mass surveillance data can too easily fall into the wrong hands and create even more significant concerns for modern society.

Installing surveillance systems in public places can be extremely helpful to ensure public safety and protect public property. At the same time, these systems can also abuse people's privacy and personal rights, and so any discussion of surveillance merits a detailed examination of the balance of personal privacy and public safety. Figure 13.2 shows the taxonomy of surveillance technologies.

**SOMEBODY'S WATCHING: HACKERS BREACH
RING HOME SECURITY CAMERAS**

In December 2019, the New York Times reported that after a couple installed a Ring security camera to the wall of the bedroom of their three daughters, a built-in speaker started piping the song "Tiptoe Through the Tulips" into the empty bedroom, according to the footage. When the couple's 8-year-old

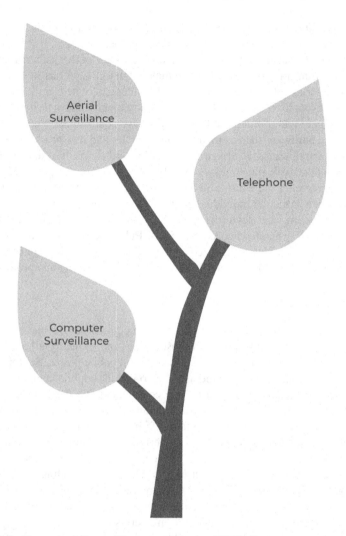

FIGURE 13.2   Taxonomy of surveillance technologies (CALEA).

daughter listened to the music and turned on the lights, a man started speaking to her, repeatedly calling her a racial slur and saying he was Santa Claus. She screamed for her mother.

The family's Ring security system had been hacked. The intrusion was part of a recent spate of breaches involving Ring, which is owned by Amazon. Ring users can monitor cameras on the company's smartphone app and speak to people inside their home and at their front door using a two-way audio feature. Hackers can gain access to these devices through usernames and passwords. Since the same username and password is reused on multiple services, it's possible for a hacker to gain access to many accounts if they do not use multi-factor authentication, which requires users to verify their identity by entering a code that they receive as a text message or by using an authentication application, in addition to their password [23].

CCTV TECHNICIAN SPIES ON HUNDREDS OF CUSTOMERS DURING INTIMATE MOMENTS

On January 23, 2021 ZDNET reported that a Texas-based CCTV technician pleaded guilty to illegally accessing the security cameras of hundreds of families to watch people in their homes get naked and engage in sexual activities. The ADT support technician's job involved installing home video surveillance cameras at customers' premises and configuring the devices to work with the company's proprietary ADT Pulse app. The technician strayed from company policy and started adding his personal email address to customers' ADT Pulse app during the installation and testing process [24].

14 Insider Threat Protection

14.1 INTRODUCTION

As the term indicates, an insider threat is an attack coming from within an organization, community, or group. The attackers, employees, or partners who have legitimate access to a system might misuse their privileges deliberately. The motivation of the attackers might be to sabotage an organization, steal intellectual property (IP), conduct industrial espionage, fraud by modification, addition, or deletion of an organization's data for personal gain, theft of information that leads to an identity crime (e.g., identity theft or credit card fraud) and, sometimes, revenge of an angry employee.

Detecting insider threats is challenging because, according to a survey [1]:

- 67% of respondents said that insiders already have credentialed access to the network and services.
- 53% said that there is an increased use of applications with potential to leak data (i.e., web, email, cloud data stores, and social media).
- 46% said that increased amount of data that leaves a protected boundary/ perimeter.
- 33% said that more end user devices are capable of theft (33%).

This chapter will review various ways of insider threat detection technologies, explain how the technologies work, the advantages and disadvantages, and the different products which use these technologies.

14.2 BRIEF HISTORICAL BACKGROUND

Before the digital age, insider threat cases were common scenarios prevalent in hundreds of movies. For example, a law enforcement agent penetrates a criminal organization and collects data or spies get inside different types of organizations to obtain data. In the digital age, insider threats became more common and more dangerous since in some cases they could facilitate the downfall of a large company or cause a huge financial or reputational loss to an organization.

In more recent history, several critical paramount cases brought focus to the importance of insider threats. In December 2006, Mike Yu accepted a job at the China branch of a US company. On the eve of his departure from Ford (where he had worked for several years previously) and before he told the automaker of his new job, Yu copied some 4,000 Ford documents onto an external hard drive, including sensitive

DOI: 10.1201/9781003038429-14

Ford design documents about engine-transmission and electric power supply systems. Yu was sentenced to 70 months in federal prison in the United States for stealing a large number of secret documents from the company [2].

Another case involved Kexue Huang, who worked at a Dow Chemical Co subsidiary from 2003 to 2008 in Indiana. He led a team of scientists developing organic insecticides and later worked for another agribusiness giant, privately held Cargill Inc. Despite signing a confidentiality agreement, Huang admitted that he passed numerous secrets about Dow's products to other researchers in Germany and China, according to his plea agreement filed in the federal court in Indiana [3].

In 2000, hackers accessed everything within Nortel Network Corporation IT systems and the information downloaded from Nortel's computers included technical papers, research and development reports, business plans, and employee emails [4].

The 2010 case of Michael Mitchell was a typical example of an insider threat from a disgruntled employee. After being terminated for his poor performance at DuPont, Mitchell, a former engineer and salesman for DuPont, kept numerous DuPont computer files and proprietary information, which he eventually used when he accepted a consultancy position at DuPont's Korean competitor [5].

The case of Edward Snowden is one of the better known cases of insider threats. In 2013, Snowden flew to Hong Kong after leaving his job as a contractor at an NSA facility in Hawaii and revealed thousands of classified NSA documents to journalists Glenn Greenwald, Laura Poitras, and Ewen MacAskill [6].

In 2015, Sergey Aleynikov, a former Goldman Sachs programmer, stole the secret source code of the company's high-frequency trading platform. During his last few days at the company, he transferred 32 MB of proprietary computer code. The company detected the case through anomalies in its network monitoring system [7].

Finally, in 2018, a disgruntled employee at Tesla company managed to create false usernames in order to make direct changes to the Tesla Manufacturing Operating System's (MOS) source code. The malicious insider also managed to export large amounts of highly sensitive data to unknown third parties.

This insider threat case could have been avoided, but the electric car company neglected to put limits on the level of privileged access given to its employees. This case drew attention to the importance of having a security program that could monitor its users' behavior to detect and prevent security incidents [8, 9].

According to Tesla CEO Elon Musk, the saboteur used his insider access to make "direct code changes to the Tesla Manufacturing Operating System under false usernames and exporting large amounts of highly sensitive Tesla data to unknown third parties" [9].

These are just a few of the more important and well-documented cases, but there are many smaller cases that are never reported. Generally, companies do not like to publicly reveal insider threat cases, since they can damage their respective reputation. With the extent of damage from potential insider threats, it is necessary for new technological solutions to address and find solutions so as to protect or minimize these threats.

14.3   HOW INSIDER THREAT PROTECTION TECHNOLOGIES WORK

ANALOGY

Imagine you own a grocery store, and you are concerned about product theft. You decide to monitor the store to ensure against any potential theft. For example, all purchases are registered so that there is a record of any product sold that leaves the store. In addition, you monitor everyone who enters and leaves the store, requiring workers to check IDs when selling alcoholic beverages and tobacco, and take on other measures. In order to do this, you hire someone to monitor the store through a two-way mirror and to report back any activity they find that could indicate theft (Figure 14.1).

Insider threat detection is simply a technology that monitors all people inside an organization who have any sort of access within that organization. This includes where people go, what files they have access or download, what emails they send and to whom they send to, etc. The technologies that do the monitoring might use different tools, but their objectives are the same. This monitoring will be used for threat detection or for forensic purposes to find evidence during the course of an investigation.

The technologies are generally quite simple. All online activities – from surfing the web to downloading and opening files to email exchanges – are collected and if they violate the company's IT policies, they alert a security agent. For example, if employees downloaded a number of files that are not needed for their jobs, then that activity might be considered suspicious and could be flagged.

Traditionally, the technologies used to prevent insider threats would work independently, such as by accessing restricted areas in a company, video monitoring, etc. Now, with artificial intelligence (AI) and data science, it is possible to create more accurate profiling of all suspicious behaviors. AI analyzes individuals or work groups daily to identify behavioral anomalies. The AI system consolidates the risk score for each user and alerts a security agent. This works similarly to a banking system that monitors users' credit card interactions and flags suspicious activity, e.g., if a credit card owner who lives in one area suddenly buys gas at a station thousand miles away from their usual location.

14.4   INSIDER THREAT DETECTION TECHNOLOGIES

Insider threat detection technologies are monitoring and reporting tools. Technologies such as Data Loss Prevention (DLP) monitor abnormal access to sensitive data such as databases or information retrieval systems. Email security technologies monitor abnormal email exchanges, particularly outbound emails. Privileged access management (PAM) monitors and controls the people with higher privileged access to data, accounts, processes, systems and IT environments. With advances in machine learning, user behavior and activity are analyzed and monitored. Thus, every unusual

FIGURE 14.1 Insider threat.

activity, exchange, or access is monitored and flagged. Below are some details about the leading technologies.

14.4.1 DATA LOSS PREVENTION

The main objective of DLP is to protect personal information and intellectual property (IP), comply with laws and regulations such as Personally Identifiable Information (PII), GDPR (for personal data of European residents) or payment card information (PCI). All these might be achieved through different tools and procedures put in one or multiple software products.

Another objective of DLP is to provide visibility of the movement of data and monitoring activities. The products that are used for DLP might be different but they all might use a policy engine with prebuilt policies while still allowing the IT department to create customized policies.

14.4.2 EMAIL SECURITY APPLICATIONS

Email security applications monitor and record all communications of employees, contractors, or people who use company email, instant messaging or voice messaging systems. The company's email servers are generally configured to create daily backups of all user data. As a result, employers can even monitor deleted or archived messages as well as the information that employees share on social networks.

Email security technologies might have the capabilities to classify corporate emails by using text mining and machine learning techniques and scrutinizing the contents of emails and preventing sensitive information from being leaked based on the email's security labels. Thus, if all emails are examined before being sent externally, then an email might be suspended and prevented from being sent out and could then be reported to a corresponding manager [10].

14.4.3 PRIVILEGED ACCESS MANAGEMENT (PAM)

One of the most important countermeasures to protect systems from any types of cyber-attacks is access management. Managing privileged access is very critical. Privileges are given to superusers enabling them to override or bypass certain security restraints. Privileges can also include permissions to perform critical actions such as shutting down systems, loading device drivers, configuring networks or systems or provisioning and configuring accounts and cloud instances. Because privileged access accounts have elevated capabilities, thus privileged users and accounts pose considerably higher risks than non-privileged accounts and users.

14.5 USER ACTIVITY MONITORING AND BEHAVIOR ANALYTICS

Technologies that are used to monitor and also be analyzed are very diverse and are evolving with the development of machine learning. The main components of user monitoring technologies are:

- Email monitoring: As mentioned before, monitoring and recording all email communications of employees.
- Activity recording: Capturing users' actions and creating an activity log. This system also captures their screen and creates a video of users' activities and searchable activity logs that can be viewed on a company system. Browsing history enables the ability to track users' Internet activities, browsing sites, most frequently visited websites, and interests.
- User behavior analytics: Uses machine learning and detects unwanted behavior.

14.6   ADVANTAGES AND DISADVANTAGES OF INSIDER THREAT DETECTION

14.6.1   ADVANTAGES

The main advantages of insider threat detection include:

- Detection of abnormal users' behavior and users with a high-risk profile
- Creating an efficient user profile and identification of users' behavior patterns
- Forensic analysis for detecting and documenting the course, reasons, culprits, and consequences of a security incident or violation of company policies
- Detection of suspicious user activity before an incident happens
- Ensuring that employees do not voluntarily or by negligence distribute or share the company's confidential information

14.6.2   DISADVANTAGES

The main disadvantages of insider threat detection technologies include:

1. All types of monitoring or surveillance systems can have a psychological impact on people and their behavior. Employees can feel vulnerable if they know they are being watched all the time.
2. False alarms due to misjudgments that can have massive consequences.
3. The cost of investigations.
4. Any level of monitoring can accumulate large amounts of data that could require significant human resources to investigate or monitor.

14.7   CONCLUSION

Various technological solutions are available to protect systems from insider threats, but it is not a simple task. Machine learning and behavior analysis can help to some degree. However, due to the unpredictability and complexity of human behaviors, most systems remain vulnerable to some degree. Figure 14.2 shows the taxonomy of insider threat protection technologies.

TESLA EMPLOYEE PREVENTED A RANSOMWARE PLOT TO STEAL COMPANY DATA FROM FACTORY

In August 2020, a man named Egor Igorevich Kriuchkov offered a Tesla employee $500,000 to install malware on Tesla's network. With that malware installed, attackers would then steal data from Tesla and hold it for ransom.

This was a very simple and effective tactic that had previously led to success. In fact, Kriuchkov claimed that another person used an insider threat at a different company and still had not been caught after several years. The Tesla

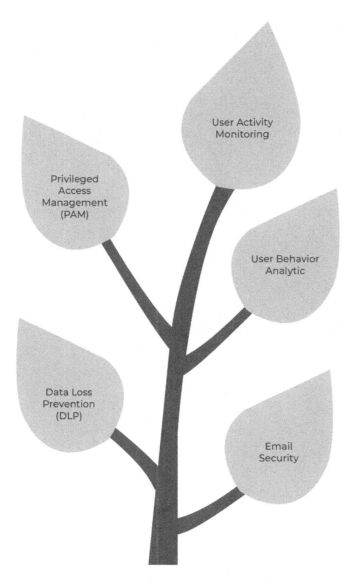

FIGURE 14.2   Taxonomy of insider threat protection technologies.

employee demonstrated a successful thwart to an insider threat, but likely not for the right reasons.

If this Tesla employee did in fact have the capability to single-handedly jeopardize sensitive company data, then this accurately demonstrates the power of a single insider threat and the damage it can cause to a company.

Tesla has been regarded as being very lucky in this situation, helping this story serve as a tale of the importance of insider threats [11].

15 Intrusion Detection

15.1 INTRODUCTION

Intrusion detection systems (IDS) monitor operating system files, network traffic, and events and analyze the data gathered to detect signs of intrusion, detect malicious activities and alert system administrators, and, in some cases, initiate action. IDS can be an appliance or software. There are a variety of ways that attackers might attack. For example, attackers might try flooding or overloading a network, gathering data about the network or finding vulnerabilities to take control of the network. Consequently, intrusion detection tools can prevent attackers from getting into the system.

15.2 BRIEF HISTORICAL BACKGROUND

IDS technologies are a relatively recent phenomenon. During the 1970s and 1980s, research was conducted to find ways to improve computer security and auditing systems.

In the 1980s, Dorothy Denning and Peter Neumann researched and developed the prototype model of an IDS, named the Intrusion Detection Expert System (IDES). IDES was a rule-based expert system trained to detect known malicious activity. During the 1990s, US government-funded research projects like Discovery, Haystack, Multics Intrusion Detection and Alerting System (MIDAS), Network Audit Director and Intrusion Reporter (NADIR) were all developed to detect intrusions [1].

During the late 1990s, a few commercial products focusing on IDS came to the market (e.g., Wheelgroup's Netranger and RealSecure). After the year 2000, more and more IDS technologies developed and came to market to analyze, alert, and scan systems very fast or to combine with other systems.

Today, IDS is an essential part of cybersecurity technologies. With a huge amount of data collected every second and the increasing complexity of network activities, monitoring of systems is extremely challenging and the need for IDS is essential.

In the last few years, the IDS field has grown considerably, and a large number of IDS programs have been developed to address specific types of intrusions.

15.3 HOW INTRUSION DETECTION TECHNOLOGIES WORK

ANALOGY

To better understand intrusion detection, consider a country that has several security agencies. In general, these agencies mostly work independently from each other. However, there is one oversight agency which constantly monitors

DOI: 10.1201/9781003038429-15

the work of all the other agencies, silently inspecting all the files, networking, and communication and alerting any anomalies and suspicious patterns to the top country official in charge. This oversight agency works in a similar way as intrusion detection does (Figure 15.1).

FIGURE 15.1   Intrusion.

The goal of IDS technologies is to act as a security agent to make sure that an organization is secure. Thus, IDS tries to identify all true attacks and identify all non-attacks. IDS technologies might be used for different purposes such as law enforcement, tracking, tracing, and prosecution of intruders. IDS can also be used as a mechanism for protecting computing resources and in identifying and correcting vulnerabilities [2].

To better understand IDS technologies, it is important to understand the objectives of security agents. "The goal of intrusion detection is to identify all true attacks and identify all non-attacks. The motivation for using intrusion detection technology may vary for different sites. Some may be interested in law enforcement, including the tracking, tracing, and prosecution of intruders. Some may use intrusion detection as a mechanism for protecting computing resources, while others may be more interested in identifying and correcting vulnerabilities." [2]

When it comes to intrusions, an IT professional might want to know the following:

- What happened?
- Who has been affected?
- How are their systems affected?
- Who are the intruders?
- Where and when did the intrusion originate and occur?
- How did the intrusion happen?
- Why did the intrusion happen?

15.3.1 NETWORK-BASED INTRUSION PREVENTION SYSTEM

Network-based intrusion prevention system (NIPS) works based on signature matching, which compares an occurring situation to previously known attacks to help assess the severity and specifics of a current attack. Like the antivirus application, the library of known attacks needs to be constantly updated to detect upcoming known attacks.

15.3.2 HOST INTRUSION DETECTION SYSTEMS

The majority, if not all, applications on each device or network create a log for each event or activity. Host intrusion detection systems (HIDS) monitors all software event logs and technically monitors the log files generated by these applications. As such, it creates a historical record of activities and functions allowing a quick search of log files for anomalies and signs of an intrusion. The detection can be done based on system administration policies, known attacks or rules that can indicate potential malicious behavior.

The HIDS runs on all devices in a network with access to the Internet and other parts of the enterprise network. HIDS has some advantages over NIPS, due to its ability to look more closely at internal traffic, as well as to work as a second line of defense against malicious packets that NIPS has failed to detect.

HIDS looks at the entire system's file set and compares it to its previous "snapshots" of those file sets. It then looks at whether there are significant differences outside normal business use and alerts the administrator whether there are any missing or significantly altered files or settings. It primarily uses host-based actions such as application use and files, file access across the system and kernel logs.

Network and host-based IDS are the most common ways of expressing this classification.

15.4 INTRUSION DETECTION TECHNOLOGIES

IDS technologies have evolved from research-based applications to identify intrusion patterns. The main components of an IDS system are monitoring detection and prevention.

There are two types of detection: knowledge-based and pattern-based detection, which constantly monitor systems and networks without human intervention.

Knowledge-based IDS detects threats and vulnerabilities by cross-referencing the threat or vulnerability signatures with a database of previous attack profiles and known system vulnerabilities to identify active intrusion attempts. Knowledge-based IDS are the most commonly used type of detection [3–5].

Pattern-based detection IDS efficiently utilizes machine learning to dynamically monitor systems and analyze network activities for any suspicious activities.

15.5 ADVANTAGES AND DISADVANTAGES OF INTRUSION DETECTION SYSTEMS

15.5.1 ADVANTAGES

15.5.1.1 Network-Based Intrusion Prevention System

The advantages of NIPS can be summarized as follows

- Suitable for large- and medium-sized organizations due to their volume of data and resources.
- When NIPS is running, it will not interrupt the network since it is running in passive mode.
- NIPS is not susceptible to direct attacks.

15.5.1.2 Host Intrusion Detection Systems

The main advantages of HIDS are the detection of attacks on a computing system, analyzing network traffic on its network interface and behavior and suggested actions that include:

- Detect unusual usage patterns from an anomaly.
- Alert anomalies to a system administrator.
- Monitor system configuration databases, registries, and figuration files.
- Alert systems and monitor in the case of file attributes changing, new file creation, or existing file deletions.
- Can be used with switched networks.
- Detect inconsistencies in the application.

15.5.2   DISADVANTAGES

15.5.2.1   Network-Based Intrusion Prevention System

- Might fail to recognize an attack when network volume becomes overwhelming.
- Some networks cannot provide all the data for analysis to NIPS due to switch limitation or lack of monitoring port capability.
- NIPS cannot analyze encrypted packets, making some of the traffic invisible to the process, and reducing its effectiveness.
- Attacks involving fragmented or malformed packets cannot easily be detected.

15.5.2.2   Host Intrusion Detection Systems

The main disadvantage of HIDS is that it does not provide prevention. The other disadvantages of HIDS include:

- Does not prevent intrusions or attacks.
- Cannot monitor at the network level.
- Does not filter incoming/outgoing traffic.
- It requires installation, configuration, and management that can be labor intensive.
- Direct attacks and attacks against the Host operating system result in compromise and/or loss in functionality of HIDS.
- Is susceptible to some DoS-related attacks.
- Target Host OS level audit logs occupy large amounts of disk space and disk capacity, thus reducing system performance.
- HIDS cannot scan/detect multi-Host and non-Host network devices.

15.6   WHICH PRODUCTS TO USE FOR INTRUSION DETECTION

There are many products that offer IDS and the following are some of the main examples: Palo Alto Networks Threat Protection, Palo Alto Networks URL Filtering PAN-DB, Cisco Firepower NGIPS (formerly Sourcefire 3D), Intrusion Detection, part of Alert Logic Professional, and Proofpoint Advanced Threat Protection.

15.7   CONCLUSION

An IDS is an overall security solution technology that monitors and detects all known anomalies and predicts suspicious activities by analyzing the pattern of networks and employing various detection techniques. IDS software tools detect computer attacks or illegitimate access and alert an IT Administrator of the detection or security breach.

IDS monitor, detect, and respond to any unauthorized activity. Figure 15.2 shows the taxonomy of intrusion detection technologies.

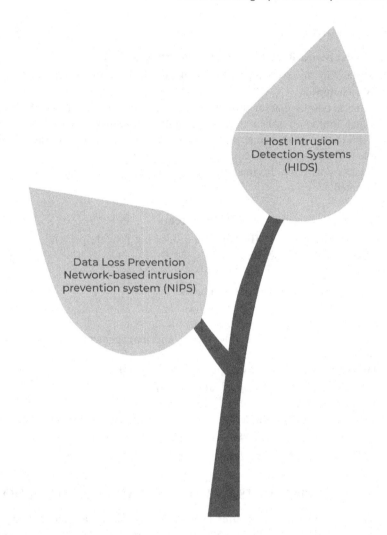

FIGURE 15.2 Taxonomy of intrusion detection technologies.

BARNES & NOBLE HIT WITH SECURITY BREACH

In 2020, the US bookseller Barnes & Noble was the victim of an intrusion, leading to unauthorized and unlawful access to certain Barnes & Noble corporate systems. Customer email addresses, billing and shipping addresses, telephone numbers, and transaction histories were all exposed during the breach [6].

It appeared that Barnes & Noble's data breach was a ransomware payload. According to reports, the company was using Pulse Secure VPN servers with an unpatched vulnerability, allowing hackers to steal usernames and passwords to infiltrate corporate systems, install ransomware, and exfiltrate data [7].

ZDNET reported that a ransomware group (Egregor) leaked data in a post on the Dark Web and claimed it belonged to Barnes & Noble customers [8].

In an earlier report, a Russian-speaking hacker forum was selling over 900 plaintext login details associated with the Pulse Secure VPN vulnerability, which included Barnes & Noble's login credentials.

According to some experts, the breach was from a phishing attack and the result of an internal staff member clicking and activating an executable file, which gave the malware an entry point.

16 Vulnerability Scanning

16.1 INTRODUCTION

Vulnerability scanning (VS) is a technology that scans a network and computer system to identify known vulnerabilities and generates risk exposure reports. VS can be run on network devices such as firewalls, routers, switches, servers, and applications in order to find potential vulnerabilities. VS is run by system administrators and IT professionals with product knowledge of specific assets. Vulnerability scanners are automated tools that check to see if networks, systems, and applications have security weaknesses that could expose them to attacks.

VS is a common practice across enterprises. Contrary to penetration testing (see Chapter 17) that is conducted by experts, VS consists of automated tools. Penetration testers and attackers alike might use VS to identify the vulnerabilities of a system to plan an attack. VS is often mandated by industry standards and government regulations to improve an organization's security posture by using several tools to identify, classify, and mitigate vulnerabilities.

After a brief history of VS, this chapter will review the main tools and technologies used in VS and review how they work.

16.2 BRIEF HISTORICAL BACKGROUND

The history of VS is very recent as it only began in the early 2000s. With more and more reliance on computer systems, the Internet, and usage of hundreds of applications by each organization, the number of vulnerabilities of the network and applications rises. The overall number of new vulnerabilities in 2019 (consisted of 20,362 cases) increased by 17.6% compared to 2018 (17,308) and by 44.5% compared to 2017 (14,086) [1].

The increase of hacking has led to security guidelines requiring reports and automated systems to extensively detect known vulnerabilities [2]. Consequently, automation to detect vulnerabilities has gradually increased with several technological solutions.

VS has improved significantly. While early programs simply checked system configuration files and file permissions, today's scanners cover several operating systems and track thousands of vulnerabilities with ease.

16.3 HOW VULNERABILITY SCANNING TECHNOLOGIES WORK

ANALOGY

To better understand VS, consider the following analogy. Imagine you have a house and you want to make sure it is secure. You might create a list of

DOI: 10.1201/9781003038429-16

the possible ways that somebody could break into your house. (For example, someone could pick the lock on your door, so you would consider the types of tools available to pick or break the lock.) You might systematically test every aspect of your house for vulnerabilities. The bigger the house, the more sophisticated tools and equipment may be needed. Thus, an assessment might be required to check each door, window, and every other vulnerable part of your house.

This same concept applies to vulnerability scanners that need to consistently check networks and electronic data, and any other possible known ways attackers might use to gain access. In the case of a home, somebody will need to check the doors and windows manually. Different software systems, however, will be used to perform these tasks on a network automatically. In addition, for homeowner's insurance, the insurance agency may require you to have tested security systems in place before agreeing to provide insurance. Similarly, with VS, some regular testing will be required to comply with certain standards (Figure 16.1).

VS uses different tools and technologies to inspect possible attacks. All vulnerability scanners create a bottom-line report of possible vulnerabilities with a lot of false positives and negatives, which result in investigations by human agents which can be costly. Consequently, expert human agents need to evaluate and define each priority to resolve all identified risks immediately and effectively.

16.3.1 PUBLIC VS APPLICATIONS

Public VS works like a search engine. The engine dynamically crawls the Internet and collects data. Then, users can run a search query to view the information about different attacks, particularly of IoT (Internet of Things) devices. The search query can be very simple or more complex, and different VS applications support utilities that detect and remove specific viruses.

16.3.1.1 System Weakness Scanning

System weakness scanning (SWS) detects a network to catch and exploit security gaps as it scans the attack surface. Different technologies exist based on the detection desired. For example, some software programs check to see if all applications and firmware are updated.

16.3.1.2 Vulnerability Classification

After the vulnerability detection, issues need to be compared with known vulnerability risks, including the National Vulnerability Database and Common Vulnerabilities and Exposures.

One of the most important parts of this is the prioritization of the issues detected.

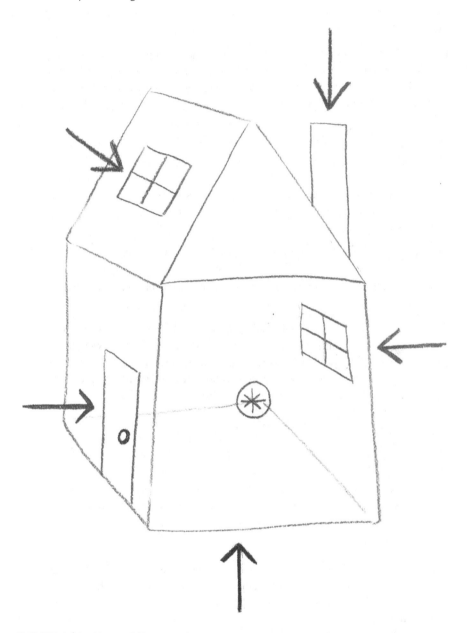

FIGURE 16.1 Vulnerability scanning.

16.3.2 WEB VS APPLICATIONS

There are a variety of WEB VS applications. Some are open source, and others are commercial license products. Depending on the type of test, the objectives, and the complexity of the network, different types of technologies might be needed.

16.3.3 INTERNAL VS TOOLS

Internal VS technologies are very diverse. In general, depending on the complexity and extent of a network in an organization, the scope of technologies that need to be used might vary significantly. This diversity explains why there are so many tools on the market. The following is a list of the main vulnerability checking tools:

- Unknown devices
- Missing updates
- OS misconfigurations
- Firewall
- Wen configuration: HTTP misconfigurations, expired SSL/TLS certificates
- Third-party applications security vulnerabilities
- Privileged Access
- File System Monitoring
- File Integrity Monitoring

Below is a list of the main open-source technologies that are available:

- OpenVAS (openvas.org)
  - OpenVAS is an open-source vulnerability scanner. Its capabilities include unauthenticated testing, authenticated testing, various high-level and low-level Internet and industrial protocols, performance tuning for large-scale scans and a powerful internal programming language to implement and type of vulnerability testing.
- Nexpose Community
  - Nexpose is a vulnerability that supports the entire vulnerability management lifecycle, including discovery, detection, verification, risk classification, impact analysis, reporting, and mitigation.
- Nikto
  - Nikto is an Open-Source General Public License (GPL) web server scanner that performs tests against web servers for multiple items, including potentially dangerous files/programs, and checks for outdated versions of servers.
- Retina
  - Retina Network Security Scanner is a vulnerability scanner used to identify, prioritize, and remediate vulnerabilities such as missing patches and configuration weaknesses.
- Wireshark
  - Wireshark is a network protocol analyzer that analyzes network traffic in real time and at a microscopic level. It is also used for troubleshooting issues on a network.

16.4 VULNERABILITY SCANNING TECHNOLOGIES

Scanners use predefined tests to identify vulnerabilities. If a vulnerability has not been publicly disclosed or the scanner's developers are unaware of it, VS will not be able to detect it. There are three groups of VS technologies: Public, web, and Internal VS.

16.4.1   Public VS

Public VS applications are specific search engines that automatically scan the Internet and collect data and share results to the public for free, with more advanced services offered for an additional subscription. For example, a search engine first scans the IPv4 (the fourth version of the Internet Protocol), public IP address range and ports, and reports the results.

Such information can later be used for benign or malicious purposes. In the case of potential attackers, they can gain reconnaissance data without directly contacting the targeted device. Additionally, IT professionals can rapidly acquire an extensive list of potential victims sharing the same vulnerability.

Several network VS tools are available, such as Shodan, Censys, Thingful, PunkSPIDER, and Zoomeye. Each tool has advantages and disadvantages [3, 4].

16.4.2   Web VS

Web VS applications are programs that run on the Internet from outside an organization's network. Web VS tests web applications for existing vulnerabilities and returns its findings as a vulnerability report to the user. The objectives are to detect vulnerabilities such as open ports in a network firewall and all possible known areas that attackers might target. Web VS applies various technologies like server-side scripting languages such as PHP or client-side scripting languages like JavaScript.

The attributes of VS detection include:

- Injection Vulnerabilities: When an attacker is able to include malicious code via insufficiently sanitized user input into an existing script that is eventually executed.
- Cross-Site Request Forgery (CSRF) vulnerabilities: Allowing an attacker to trick an authenticated victim into submitting specifically crafted requests that trigger unintended actions from the victim. Password resets, financial transactions, and other state-changing actions are examples of functions that have to be protected from CSRF attacks.
- Reflected Cross-Site Scripting vulnerabilities: When a web application processes the parameters of an HTTP request improperly.
- Cross-Site Scripting: A web security vulnerability that allows an attacker to induce users to perform actions that they do not intend to perform.
- DOM-based XSS: In a DOM-XSS vulnerability, the final malicious code is injected by the victim's web browser and not by the server [5].

16.4.3   Internal VS

Internal organization VS tools are used inside an organization to detect vulnerabilities at different locations in a network, applications, operating systems, etc. There are various tools that might be used for a specific type of vulnerability based on the objective – for example, network vulnerability in the case of insider threats. Therefore, scanning can occur in two different ways: unauthenticated and authenticated. In unauthenticated scanning, the method is similar to external VS as it looks at vulnerabilities without any authentication access privileges. On the contrary, authenticated scanners allow for direct network access using remote protocols such as secure shell (SSH) or remote desktop protocol (RDP).

The objective of internal VS is to detect accounts with weak passwords, network configurations, etc. [6–8].

16.5 ADVANTAGES AND DISADVANTAGES OF VULNERABILITY SCANNING

16.5.1 ADVANTAGES

- VS detects vulnerabilities before attackers discover them.
- Internal VS helps to mitigate risk for each different profile with privileged access.
- Internal VS helps to understand security holes from the inside. This would help, for example, external VS to understand how attackers might approach.

16.5.2 DISADVANTAGES

- High cost
- Requires expert human resources familiar with the technology
- Often inaccessible to smaller organizations with limited financial and human resources

16.5.3 CONCLUSION

With the continuing rise of cyber-attacks all over the world, protecting computer systems is paramount to all other aspects of system design. Frequent VS is a fundamental preventive measure that can make an organization aware of potential risks and take preventive measures.

Since attackers spend significant time and resources to look for vulnerabilities of computer systems, it's imperative for enterprises to do the same in order to prevent such attacks and VS is a good tool to do that.

Due to the importance and high cost of dynamic detection, automation of known issues can offer a cost-effective approach. There are many security tools and open-source solutions for the network administrator to choose from based on their needs and risks. Figure 16.2 shows the taxonomy of VS technologies.

100 MILLION CAPITAL ONE CUSTOMERS' PERSONAL INFO IS HACKED

In March 2019, a hacker gained access to more than 100 million Capital One customers' accounts and credit card applications. The breach affected around 100 million people in the United States and about 6 million people in Canada, according to Capital One.

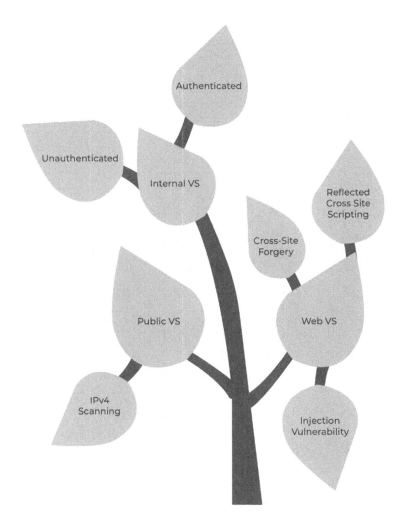

FIGURE 16.2   Taxonomy of vulnerability scanning technologies.

Eventually, Paige Thompson was arrested for the crime, and she said that she gained access to the Capital One system by using a special command to extract files in a directory stored on Amazon's servers [9].

SEVERE VULNERABILITY IN APPLE FACETIME

In January 2019, it was discovered that Apple's FaceTime video was allowing people to secretly eavesdrop on other users before they answered calls. This bug affected any Apple iPhone, iPad, or Mac that could run FaceTime [10].

17 Penetration Testing

17.1 INTRODUCTION

A penetration test (pen test) is the process of allowing authorized experts to hack a computer system in order to discover the vulnerabilities of a system and to remove those areas of weakness. A pen test is not a technology in itself, because it authorizes human experts to use all possible ways to hack a computer system. Testers might use a variety of different tools, such as vulnerability scanning (Chapter 16), intrusion tools (Chapter 15), or social engineering techniques.

Penetration testing is considered a formal procedure to discover security vulnerabilities, flaw risks, and an unreliable environment. Some people differentiate between a pen tester and "ethical hacking." The difference might be due to the fact that a pen tester uses known protocols to test a system (e.g., Quality Testing in software engineering), but ethical hackers are authorized to use all means to hack a system, which could include social engineering techniques.

However, since the term "ethical hacking" is used in a positive way, I will be using penetration testing for both approaches. Additionally, although some people use the terms "white" or "black" testing, I will refrain from using these due to the potential racial meaning or implications.

After a brief history of pen testing, this chapter will review the main tools and technology used in pen testing to hack a computer system [1].

17.2 BRIEF HISTORICAL BACKGROUND

During the 1960s, pen testing originated by the RAND Corporation of the US government to identify vulnerabilities in systems, networks, hardware, and software. Gradually businesses started to create pen tester teams to find vulnerabilities in computer networks and systems to protect them from real hackers.

In 1972, James P. Anderson outlined a series of definitive steps in a report for the US government [2]. Anderson's approach included identifying vulnerabilities and designing an attack to exploit them, then finding the weakness in that attack itself and the different ways to neutralize its threat. Anderson is now considered one of the early pioneers in penetration testing.

In 1974, the US Air Force conducted an ethical hack on its Multics system. It was one of the earliest known white hat testing in the United States, where plenty of vulnerabilities were revealed. White hat testing is when a computer security expert conducts pen testing. During the 1980s and 1990s, several systems were developed and used for pen testing by a broad range of government, military, and corporate entities.

With the growing number of cyber-attacks on government and private enterprise computer systems, pen testing has gradually become one of the main ways to

DOI: 10.1201/9781003038429-17

simulate a system's vulnerabilities and security procedures. Some famous hackers became security experts in penetration testing.

17.3   HOW PENETRATION TESTING TECHNOLOGIES WORK

ANALOGY

Imagine you have built a home with all the strongest possible security systems in place including sophisticated locks, surveillance cameras, reinforced doors, etc. Then, you hide something in your home, in a location that only you know. To make sure it is well hidden, you ask a few professional experts in home security, people who are familiar with the many methods of burglary, to find what you have hidden. These hired experts use all the techniques at their disposal to get into the home, look for the object, find it and get it out, all while being noticed as little as possible.

Once this is accomplished, these experts can show you how they were able to gain access and find the object, so you can remove those vulnerabilities. For instance, perhaps you hid the object under your bed, and they inform you that most people hide important objects under their beds, which makes them easy to find. Now you know to hide things in a safer location (Figure 17.1).

Pen testing procedures and scope are determined based on the objective. As a result, the technologies that might be used would be different in each case. To conduct penetration testing, there are several steps that need to be followed [3]:

- Goal Settings: Determining the main objectives of the pen testing based on available resources and the extent of testing.
- Information Gathering: Collecting as much information as possible prior to performing the attack.
- Service Indentation: Identifying all open and close ports for the known vulnerabilities of the target machine.
- Vulnerability Identification: Identifying vulnerability at the OS, system, or even network level.
- Vulnerability Exploitation: Identifying where a hacker strives to retain their control over a target with backdoors, rootkits, or Trojans. Compromised machines can even be used as Bots and Zombies for further attacks.

17.4   PENETRATION TESTING TECHNOLOGIES

Fundamentally, the technology used in pen tests is not much different from the technology that is used in vulnerability scanning. Different technologies might be used depending on the type of pen testing. Pen testing can target specific areas, i.e., a network, application, operating system, etc. The test can be run from either inside or outside the organization and can be performed manually or automatically,

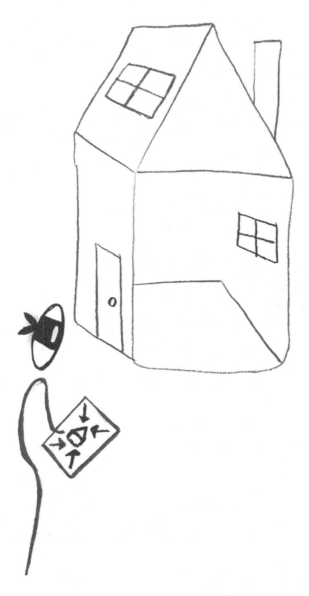

FIGURE 17.1 Penetration testing.

depending on the requirements. However, automated penetration testing is an emerging area. Automated or semi-automated systems might use various tools, which help small industries or even large-scale networks to identify security concerns and possible entry points for hackers. The extent of testing depends on available resources, and it's important to note that most pen testing uses open-source tools and technologies.

The following are the main technologies that are used for most pen testing [4–6].

Port scanners collect information about the network environment and available services to be targeted. There are several open-source tools available to help with this task.

Network Mapper (https://nmap.org/):

- Scans and discovers open ports on specific networks/hosts.
- Identifies potential hosts that are responding to network requests.
- Discovers the operating system name and version, along with network details used by the host.
- Determines what kind of apps are running and their version numbers.

Unicornscan (http://www.unicornscan.org/) is a port scanning tool for scanning servers and hosts to see which available ports are being utilized for network communications. The main features include:

- Asynchronous stateless TCP scanning
- Asynchronous UDP scanning
- IP port scanner and service detection
- Remote operating system detection
- Enabling multiple modules from command-line

Angry IP Scan (https://angryip.org/) is an open-source and cross-platform network scanner designed to scan IP addresses and ports.

Netcat (https://nmap.org/ncat/) is a network debugging and investigation tool. Its main features include:

- Built-in port-scanning capabilities
- TCP and UDP port scan support
- Verbose port scanning
- Read command line arguments from standard inputs

Zenmap is a multi-platform open-source application used to save and compare scans, view network topology maps, view displays of ports running on a host or all hosts on a network, and store scans in a searchable database. Its main features include:

- Saving scan results in a database
- Searching the results database
- Comparing current scan results with previous scans
- Saving port scan profiles for frequently used port discovery options

17.4.1 Application Scanners

An application scanner detects security weaknesses in web-based applications that include but are not limited to:

- Memory buffer overruns
- Cookie manipulations

- Malicious SQL injections
- Cross site scripting (also known as "XSS")

There are a variety of tools, open source or licensed, that are available for different tasks. Below are the primary open source tools:

Grabber is a web application scanner that detects vulnerabilities in a website. It is good for scanning small websites. The main features include:

- Cross-site scripting
- SQL injection
- Ajax testing
- File inclusion
- JS source code analyzer
- Backup file check

Vega is a free and open-source web security scanner and testing platform that helps to find and validate SQL Injection. The main features are:

- Cross-site scripting
- Stored cross-site scripting
- Blind SQL injection

Zed Attack Proxy is an open-source web application security scanner with the following features:

- Intercepting proxy
- Automatic scanner
- Traditional but powerful spiders
- Fuzzer
- Web socket support
- Plug-n-hack support
- Authentication support
- REST-based API
- Dynamic SSL certificates
- Smartcard and client digital certificates support

17.5 ADVANTAGES AND DISADVANTAGES OF PENETRATION TESTING

17.5.1 Advantages of Penetration Testing

Pen testing is a proactive approach to security and helps to investigate data breaches or network intrusions to discover any leads to the leakage of data or theft of intellectual property. Pen testing is a proactive security approach, and its advantages can be summarized as follows:

- Helps identify patterns by looking at smaller vulnerabilities that are part of a more complex attack system but that won't be of much concern if analyzed individually.
- Can find known and unknown hardware or software flaws.
- Able to exploit security vulnerabilities.
- Allows the exploration of real risks and has an accurate representation of IT infrastructure security.
- Allows vulnerabilities to be identified and fixed before they are exploited in a much more effective way than automated tools.
- Identifies patterns by looking at smaller vulnerabilities.
- Checks the efficacy of one's defensive mechanisms far beyond the depth of analysis provided by a vulnerability assessment. This allows identifying whether weaknesses are originating from human errors or from technical issues.
- Evaluates IT policies and procedures.
- Tests systems with attacks that are as close as possible to real-world incidents.

17.5.2 DISADVANTAGES OF PENETRATION TESTING

Penetration testing should not be the only method used to secure systems due to the following reasons and disadvantages:

- It is not a full security audit.
- It takes a pen tester more time to inspect a given system to identify attack vectors than doing a vulnerability assessment, meaning the test scope is greater.
- A tester's actions can be disruptive for business activities as they mimic a real attack.
- It is labor intensive and can therefore represent an increased cost, which some organizations might not be able to afford. This is especially true when an outside firm is hired to carry out the task.
- There might be a false sense of security. Being able to withstand most penetration testing attacks might give the sense that systems are 100% safe. In most cases, however, penetration testing is known to company security teams who are ready to look for signs and are prepared to defend whereas real attacks are unannounced and, above all, unexpected.

17.6 CONCLUSION

Pen testing is not a technology in itself as pen testers use several technologies to test specific vulnerabilities or conduct authorized hacking to catch an unknown vulnerability. Pen testing is recommended to be conducted regularly, however, the success of pen testing is dependent on the skill of the test team, test environment, and technologies and tools used. Although it might be conducted in a short period of time, pen testing might also be resource intensive and requires very skillful professionals and access to appropriate tools (Figure 17.2).

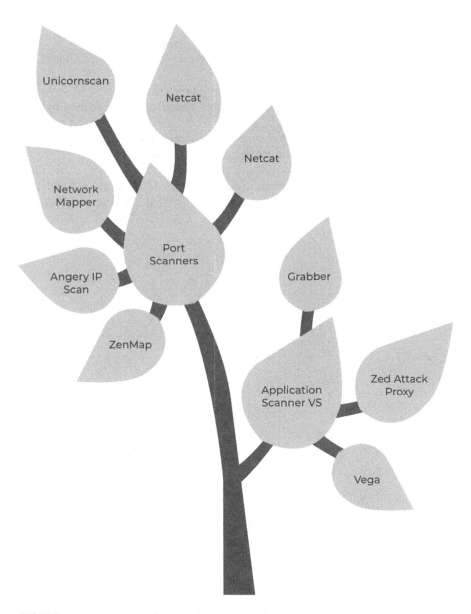

FIGURE 17.2   Taxonomy of penetration testing technologies.

WE HACKED APPLE FOR 3 MONTHS: HERE'S WHAT WE FOUND

In October 2020, Sam Curry, a Web Application Security Researcher, reported in a blog that he and four other colleagues worked together to hack the Apple bug bounty program. They found a variety of vulnerabilities in core portions of Apple infrastructure that would have allowed an attacker to:

- Fully compromise both customer and employee applications.
- Launch a worm capable of automatically taking over a victim's iCloud account.
- Retrieve source code for internal Apple projects.
- Fully compromise an industrial control warehouse software used by Apple.
- Take over the sessions of Apple employees with the capability of accessing management tools and sensitive resources.

A total of 55 vulnerabilities were discovered with 11 classified as "critical severity," 29 "high severity," 13 "medium severity," and 2 "low severity." [7].

18 Conclusion

In 2020, the number of smartphone users worldwide surpassed three billion, and the number of Internet of Things (IoT) connected devices is forecasted to reach 50 billion by 2030 around the world [1, 2]. Most adults, and a large majority of children, have a mobile phone; homes are connected to the Internet, enterprises and government agencies provide services online, and the devices will continue to be Internet enabled. All of these changes result in more cyber-attacks. These attacks don't just affect large companies or organizations, but each and every individual is at risk. Consequently, people need to look at different kinds of technologies to protect their digital assets and private information, while organizations strive to protect their assets, customer data, and any other information used in their operations and business.

We can see the signs of these changes at the personal level. For example, when one purchases a router, the manufacturer offers additional monthly services to protect their network from attacks; or when customers are invited to buy malware protection and virus detector software and services; or the proliferation of VPN services to protect communications, just to name a few.

Cyber-attackers, government surveillance agencies, police departments, investigators, and criminal organizations have all become more sophisticated in their ways of acquiring data from people and organizations. Today we already notice the growing number of industrial espionage incidents, cyberwarfare attacks, and the many ways in which private information of people is gathered through social networking to influence their opinions for political gains around the world.

All these changes require people and organizations to use different technologies to protect themselves from attacks. As a result, the cybersecurity field will grow exponentially and evolve with new solutions. There will be a growing need for professionals and technological innovation. This book is a humble effort to provide a framework and understanding of the basics of cybersecurity technologies.

DOI: 10.1201/9781003038429-18

Glossary

Access Tokens: An access token comprises the safety credentials for a login session and recognizes the user's password. It should be kept undisclosed.

Advanced Encryption Standard (AES): AES is a requirement for the encryption of electronic documents established by the US National Institute of Standards and Technology (NIST) in 2001.

AlexaReputation: Alexa uses the information provided by its toolbar operators for ranking their individual websites vis a vis another website.

Algorithms: An algorithm is a classification of well-defined, computer-implementable instructions to accomplish a computation in mathematics.

Application Proxy: Application Proxy is a feature of Azure AD that enables users to enter on-premises web applications from a distant client.

ARPANET: The Advanced Research Projects Agency Network (ARPANET) is the leading wide-area packet-switching network with spread control and one of the main networks to implement the TCP/IP protocol suite. Both skills became the practical foundation of the Internet.

Asymmetric Key: Asymmetrical encryption or public-key cryptography is a moderately unique method of encryption, as compared to symmetric encryption.

Authentication Code Protocol (CCPM): CCPM is an encryption procedure planned for Wireless LAN products that apply the standards of the IEEE 802.11i alteration to the innovative IEEE 802.11 standard.

Backoff: Backoff is a malware marks point of sale (POS) system. It is used to steal credit card information from sale machines at retail stores.

Biometrics: Biometrics are physical or behavioral human characteristics that can be utilized to digitally recognize a person to allow access to systems, devices, or data.

Blowfish (Cipher): Blowfish is a symmetric-key block cipher used to substitute DES or IDEA.

Bluetooth Low Energy (BLE): BLE is a wireless personal area network technology designed and marketed by the Bluetooth Special Interest Group (Bluetooth SIG) aimed at novel applications in the health care, fitness, beacons, security, and home entertainment industries.

Brute Force Attack: A brute force attack is the technique to advance access to a site or server (or whatever that is password endangered) by initiating numerous mixtures of usernames and passwords over and over till it gets in.

Bulk Encryption: It means encrypting a big amount of circuits after they have been multiplexed.

CCMP – Counter Mode Cipher Block Chaining Message Authentication Code Protocol (Counter Mode CBC-MAC Protocol) or CCM mode Protocol: CCMP or CCM mode protocol is an encryption protocol intended for

Wireless LAN codes that apply the standards of the IEEE 802.11i adjustment to the original IEEE 802.11 standard.

ClamAV: ClamAV® is an open-source customary for mail entry perusing clamav. net.

Communications Assistance for Law Enforcement Act (CALEA): CALEA, also known as the "Digital Telephony Act," is a US wiretapping law approved in 1994. Its aim is to improve the skill of law enforcement agencies to conduct a lawful intervention of communication by demanding that telecommunications carriers and producers of telecommunications tools adapt and design their tools, facilities, and amenities to ensure that they have built-in capabilities for targeted surveillance, allowing federal agencies to select wiretap any telephone traffic.

Compatible Time-Sharing System (CTSS): CTSS was established at the Massachusetts Institute of Technology Computation Center in the 1960s and 1970s. The project of the CTSS signifies the beginning of the impression that operating systems can operate on multiple threads or "multitask." Time-sharing means the system can assign resources for two chores or procedures at once.

Computer Emergency Response Team (CERT): A Computer Emergency Response Team (CERT) is a group of data safety experts accountable for the protection alongside, detection of and reaction to an organization's cybersecurity events.

Confidentiality Integrity and Availability (CIA): CIA triad is a model designed to direct policies for information security within an organization.

Content Injection: Content injection is the performance of operating what a user sees on a place by totaling parameters to their URL.

Content Spoofing: Content spoofing is a kind of exploit to current a faked or modified website to the operator as if it were legitimate.

Crimeware: It is a range of techniques to bargain intimate data.

Cross-Site Request Forgery (CSRF): CSRF is a web safety susceptibility that allows an attacker to persuade users to do actions that they do not mean to achieve.

Cross-Site Scripting (XSS): XSS is a client-side code injection attack. The attacker impartial is to perform malicious scripts in a web browser of the victim by including malicious code in a legitimate web page or web request.

Cryptographic Salt: Cryptographic salt is accidental data that is used as an additional input to a one-way function that hashes data, a password, or passphrase. Salts are utilized to safeguard passwords in safe system.

Curve Cryptography (ECC): ECC is a method to public-key cryptography founded on the algebraic construction of elliptic curves over finite fields.

Data Encryption Standard (DES): DES is a symmetric-key algorithm for the encryption of digital data. Due to its short key length of 56 bits, it is too insecure for new applications.

Data Mining Techniques: Data mining is a process of discovering patterns in large data sets involving methods at the intersection of machine learning, statistics, and file systems.

Data Revocation: Data revocation refers to the revocation of a digital security certificate. It is the procedure of provably deleting all copies of a data set.

DDoS Attacks: DDoS attack happens when multiple systems flood the bandwidth or resources of a targeted system, usually one or more web servers.

Document Object Model (DOM): DOM is a programming API for HTML and XML documents. It describes the logical structure of documents and the method a document is accessed and manipulated.

DOM-Based XSS: DOM-Based XSS is an XSS attack wherein the outbreak payload is executed as a result of adjusting the DOM "environment" in the victim's browser used by the innovative client side script, so that the customer side code runs in an "unexpected" manner.

Domain Name System (DNS): DNS is a naming system for computers, services, or other resources linked to the Internet or a private network. It connects numerous information with domain names allocated to each of the contributing entities.

Domain Name System (DNS) Protocols: The Domain Name System (DNS) is the phonebook of the Internet. Humans enter information online over domain names, like nytimes.com or espn.com. Web browsers relate through Internet Protocol (IP) addresses. DNS translates domain names to IP addresses so browsers can burden Internet resources.

Domain Name System-Based Blackhole Listing (DNSBL): DNSBL is/are lists of IP addresses that are supposed of sending spam. It is used to stop unwanted email messages from accomplishment of unsuspecting recipients.

Dynamic Host Configuration Protocol (DHCP): DHCP is a network management protocol used to automate the procedure of configuring devices on IP networks.

Earthmovers' Distance (EMD): EMD is an amount of the detachment between two probability deliveries over a region D. In mathematics, this is known as the Wasserstein metric.

Elliptic Curve Digital Signature Algorithm (ECDSA): Elliptic Curve Digital Signature Algorithm or ECDSA is a cryptographic algorithm used by Bitcoin to safeguard that funds can only be consumed by their owners.

Endpoint Detection and Response (EDR): EDR platforms are a class of endpoint security tools, constructed to provide endpoint visibility. EDR are used to notice and respond to cyber threats and exploits.

Enigma Code: Enigma used by the German high command to encode strategic messages before and during World War II. The Enigma code was broken by the Poles under the leadership of mathematician Marian Rejewski in the early 1930s.

European Institute of Computer Antivirus Research (EICAR): EICAR test file is a computer file that was advanced by the European Institute for Computer Antivirus Research (EICAR) and Computer Antivirus Research Organization (CARO), to test the reply of computer antivirus (AV) programs.

Federal Information Processing Standards (FIPS): FIPS standards are issued to establish requirements for various purposes such as safeguarding computer security and interoperability. Establish requirements can be used for cases in which appropriate industry standards do not already exist.

General Data Protection Regulation (GDPR) (for personal data of European residents): GDPR is a European Union regulation that needs businesses to guard the personal data and privacy of EU citizens for dealings that happen within EU member states.

Global Positioning System (GPS): GPS is a satellite-based radio navigation system owned by the US government and functioned by the US Space Force.

Google PageRank: PageRank (PR) is an algorithm used by Google Search to rank web pages in their search engine results.

Hash Table: Hash table is a data structure which stores information in an associative manner. In a hash table, data is stored in a collection format, where each data value has its own distinctive index value.

Hashing: Hashing is the alteration of a string of characters into a usually smaller fixed-length worth or key that represents the original string.

Heuristic Detection: Heuristic detection analysis is a technique of detecting viruses by observing code for suspicious properties.

Host-Based Intrusion Prevention Systems (HIPS): A host-based intrusion anticipation system (HIPS) is a program employed to defend critical computer systems containing crucial data alongside viruses and other Internet malware.

Host Intrusion Prevention System: HIPS is a software set that monitors a single host for doubtful activity. HIPS analyze events occurring within that host.

HyperText Transfer Protocol (HTTP): HTTP is an application protocol for spread, collaborative, hypermedia information systems.

Hypervisor: A hypervisor is software that makes and runs virtual machines (VMs). A hypervisor allows one host computer to care multiple guest VMs by nearly sharing its resources, such as memory and processing.

IBM's APL Network: It is a programming language for users and developers.

Identity Verification (IV): IV is a process that ensures a person's identity matches the one that is hypothetical to be.

Infrastructure-as-a-Service (IaaS): IaaS is an instant computing infrastructure, provisioned and achieved over the Internet.

Injection Vulnerabilities: Injection vulnerability allows an attacker to include malicious code via insufficiently sanitized user input into an existing script that is finally executed.

Internet Control Message Protocol (ICMP): ICMP is a supporting protocol which is used by network devices, including routers, to direct error messages and operational information indicating achievement or failure when collaborating with another IP address.

Internet Protocol (IP): IP is a protocol, or set of rules, for routing and addressing packets of data so that they can travel through networks and arrive at the correct destination.

Internet Protocol Security (IPsec): IPsec is a safe network protocol that authenticates and encrypts the packets of data to provide secure encrypted communication among two computers over an IP network.

Internet Protocol Version 6 (IPv6): IPv6 is the most recent version of the IP. It was advanced to deal with the problem of IPv4 address exhaustion.

Intrusion Detection Expert System (IDES): IDES is a system that detects user behavior on one or more monitored computer systems and flags doubtful events.

Intrusion Detection System (IDS): An intrusion detection system is a device or software request that monitors a network or systems for malicious activity or policy damages.

IPV6: Internet Protocol version 6 (IPv6) is the most recent version of the IP, the communications protocol that provides an identification and location system for computers on networks and routes traffic across the Internet.

Iris: The iris is a thin, annular construction in the eye, responsible for controlling the diameter and size of the pupil and thus the sum of light reaching the retina. Eye color is defined by that of the iris.

ISO/IEC Standards Committee: ISO/IEC is a committee of the International Organization for Standardization (ISO) and the International Electrotechnical Commission (IEC).

Key Reinstallation Attack (KRACK): KRACK is a replay attack on the Wi-Fi Protected Access protocol that safeguards Wi-Fi connections.

Keyloggers: Keyloggers are software intended to record keystrokes made by a user. The keystroke loggers record the information that user types into a website or application and sends back to a third party.

Levenshtein Distance (LD): LD is an amount of the similarity between two strings (the source string (s) and the target string (t)). The distance is the number of deletions, insertions, or substitutions obligatory to transform into.

Local Area Network (LAN): LAN is a computer network that spans a comparatively small area. Most often, a LAN is limited to a single room, building, or group of buildings.

LoRa (Long Range): LoRa (short for long range) is a spread spectrum modulation technique imitative from chirp spread spectrum (CSS) technology. Semtech's LoRa devices and wireless radio occurrence technology is a long-range, low-power wireless platform that has become the de facto technology for Internet of Things (IoT) networks worldwide.

Magic Lantern: Magic Lantern is keystroke logging software industrialized by the United States' Federal Bureau of Investigation (FBI).

Malware: Malware is a malicious software designed to cause extensive damage to data and systems or to gain unauthorized access to a network.

Man-In-The-Middle Attacks (MITM): MITM is an attack where attacker secretly relays and perhaps alters the communications between two parties who believe that they are straight communicating with each other.

Massachusetts Institute of Technology (MIT): MIT is a private research university in Cambridge, Massachusetts. The institute is a land-grant, sea-grant, and space-grant university, with a city campus that extends more than a mile (1.6 km) alongside the Charles River.

MD5/SHA Checksum: MD5 and SHA-1 Checksum Utility is a standalone freeware tool that generates and confirms cryptographic hashes in MD5 and SHA-1.

Media Access Control Layer: MAC Layer is one of two sublayers that sort up the Data Link Layer of the OSI model. The MAC layer is responsible for moving data packets to and from one Network Interface Card (NIC) to additional across a shared channel.

Morse Code: Morse code is a technique used in telecommunication to encode text fonts as standardized sequences of two different signal durations, called dots and dashes or dits and dahs. Morse code is named after Samuel Morse, an inventor of the telegraph.

The Multics Intrusion Detection and Alerting System (MIDAS): MIDAS is a rules-based system intended to perform detection of online intrusions.

Multi-Factor Authentication (MFA): MFA is a security system that confirms a user's identity by requiring multiple credentials.

National Aeronautics and Space Administration (NASA): NASA is the federal agency that is accountable for aerospace research, aeronautics, and the civilian space program.

National Institute of Standards and Technology (NIST): NIST is a physical sciences laboratory and a non-regulatory agency of the United States Department of Commerce.

Near-Field Communication (NFC): NFC is a contact-less communication technology based on a radio frequency (RF) field using a base frequency of 13.56 MHz. NFC technology is intended to exchange data between two devices through a simple touch gesture.

Network Address Translation (NAT): NAT is designed for IP address preservation and operates on a router connecting two networks together. NAT translates the private (not globally unique) addresses in the interior network into legal addresses, before packets are forwarded to another network.

Network Anomaly Detection and Intrusion Reporter (NADIR): NADIR is an expert system that provides a real-time security auditing for intrusion and misuse detection.

Network Audit: Network auditing is a process in which your network is mapped both in relations of software and hardware.

Next-Generation Antivirus Software (NGAV): NGAV takes old-style antivirus software to a new, advanced level of endpoint security protection. It goes elsewhere known file-based malware signatures and heuristics because it's a system-centric, cloud-based approach.

Norton Antivirus: Norton Antivirus is an antivirus or anti-malware software product, established and distributed by Symantec Corporation since 1991.

One-Time Passwords (OTP): A one-time PIN code is a code that is valid for only one login session or transaction using a mobile phone.

OpenAntivirus Project: OpenAntiVirus.org is an OpenSource project

OpenFlow: OpenFlow a network protocol to manage and direct traffic among routers and switches from various vendors.

Open Systems Interconnection model (OSI): OSI model is a model that characterizes and standardizes the communication functions of a telecommunication or computing system without regard to its underlying internal structure and technology.

Over the Air Firmware (OTA): An over-the-air update is the wireless delivery of new software or data to mobile devices. Wireless carriers and original equipment manufacturers (OEMs) typically use over-the-air (OTA) updates to organize firmware and configure phones for use on their networks.

Packet Encryption: Internet Protocol Security (IPsec) is a secure network protocol suite that validates and encrypts the packets of data to provide safe encrypted communication between two computers over an IP network.

Payment Card Information (PCI): The Payment Card Industry Data Security Standard (PCI DSS) is an information security standard for organizations that handle branded credit cards from the major card schemes.

Platform-as-a-Service (PaaS): PaaS is a complete development and deployment atmosphere in the cloud.

Port 80 (HTTP): On a web server or Hypertext Transfer Protocol daemon, port 80 is the port that the server "listens to" or expects to receive from a web client, presumptuous that the default was taken when the server was configured or set up. A port can be specified in the range from 0-65536 on the NCSA server. However, the server administrator configures the server so that only one port number can be recognized. By default, the port number for a web server is 80. Experimental services may sometimes be run at port 8080.

Port 443 (HTTPS): The port 442 is used for secure web browser communication. It is the standard port for all secured HTTP traffic. It is essential for most modern web activity. Encryption is essential to protect information, as it makes its way between your computer and a web server.

Pre-Sharing of Keys (PSK): PSK is a shared secret which was formerly shared between the two parties using some secure channel before it needs to be used.

Proxy Server Firewalls: A proxy firewall is a network security system that protects network resources by filtering messages at the application layer.

Pseudo-Hadamard Transform (PHT): PHT is a reversible alteration of a bit string that provides cryptographic diffusion

Quality of Service (QoS): Quality of service (QoS) refers to any technology that manages data traffic to reduce packet loss, latency, and jitter on the network.

Quick Response Code (QR): QR Code is a type of barcode that can be read simply by a digital device and which stores information as a series of pixels in a square-shaped grid.

Radio Frequency Identification (RFID): RFID uses electromagnetic fields to mechanically identify and track tags attached to objects. An RFID tag consists of a tiny radio transponder, a radio receiver and transmitter.

RAND Corporation: The RAND Corporation is a nonprofit institution that helps recover policy and decision making through research and analysis.

Random Access Memory (RAM): RAM is one of the most important elements of computing. It is the super-fast and provisional data storage space that a computer needs to access right now or in the next few moments.

Random Bit-Generation: It is an algorithm for generating a sequence of numbers whose properties estimated the properties of sequences of random numbers.

Ransomware: Ransomware is a malicious software that infects user computer and displays messages challenging a fee to be paid in order for user system to work again.

RC4 (Rivest Cipher 4): Also known as ARC4 is a stream cipher with an instinctive user interface. However, multiple vulnerabilities have been discovered in it that makes it insecure.

Reinstallation Attack (KRACK): KRACK is a replay attack on the Wi-Fi Protected Access protocol that safeguards Wi-Fi connections.

Repository-Based Software Engineering (RBSE): The Repository-based Software Engineering Program (RBSE) is described to notify and update senior NASA managers about the program.

Retina: A layer at the back of the eyeball containing cells that are subtle to light and that trigger nerve impulses that pass via the optic nerve to the brain, where a visual image is formed.

Rivest Cipher 4 (RC4): RC4 is a stream cipher.

Rootkits: A rootkit is a malicious software that allows an unauthorized user to have privileged access to a computer and to restricted areas of its software. A rootkit can cover malicious tools such as keyloggers, banking credential stealers, password stealers, antivirus disablers, and bots for DDoS attacks.

RSA (Rivest–Shamir–Adleman): RSA is a public-key cryptosystem. RSA encryption key is public and distinct from the decryption key which is kept private.

Sandbox Detection: Sandbox detection analyzes malware by executing code in a safe and isolated environment to observe that code's behavior and output activity.

Science Research Associates (SRA): Science Research Associates (SRA) is a Chicago-based publisher of informative materials and schoolroom reading comprehension products.

Secure Email Gateways (SEG): A Secure Email Gateway (SEG) is a device or software used to monitor emails that are being sent and received. All messages that are spam, phishing, malware, fraudulent according to IT policies are blocked.

Secure Sockets Layer (SSL): SSL is a standard security technology for beginning an encrypted link between a server and a client – typically a web server (website) and a browser, or a mail server and a mail client.

Security Information and Event Management (SIEM): Security Information and Event Management (SIEM) is a software solution that aggregates and analyzes activity from many different resources across your entire IT infrastructure. SIEM collects security data from network devices, servers, domain controllers, and more.

Security Keys: A security key is a small physical device that looks like a USB thumb drive and used as a key to access a computer system.

Security Operations Center (SOC): SOC is a facility that houses an information security team responsible for monitoring and analyzing.

Signature-Based Detection: Signature-based detection is a process where a single identifier is established about a known threat so that the threat can be identified in the future.

Simultaneous Authentication of Equals (SAE): SAE is a secure password-based authentication and password-authenticated key agreement process.

SMBs Trust: SMBS Trust Agreements defines the terms of the deal, describes how the deal will be executed, and lists the deal's parameters or limitations.

Smishing: Smishing is any kind of phishing that involves a text message. It is mainly scary because people tend to be more inclined to trust a text message than an email.

Software-Defined Networking (SDN): SDN is an approach to network management that enables dynamic, programmatically efficient network configuration. It improves network performance and monitoring and making it more like cloud computing than traditional network management.

Sony DRM Rootkit in 2005: Music CDs that secretly installed a rootkit on computers.

Spear Phishing: The practice of sending emails from a known or trusted sender to induce targeted individuals to disclose confidential information.

Spyware: Spyware is a software that infiltrates to a computing device, stealing the Internet usage data and sensitive information.

SQL injection: A code injection technique that destroy database.

Stingrays: Stingrays is cellphone surveillance devices that can obtain the content of voice and text communications of cellphone users in the vicinity and listen to chats by using a person's phone as a bug to listen in on conversations.

Stuxnet Worm: Stuxnet is a malicious computer worm, first uncovered in 2010. It targets supervisory control and data acquisition (SCADA) systems and is believed to be accountable for causing substantial damage to the Iranian nuclear program of Iran by the United States and Israel.

Symantec: An American consumer software company.

Symmetrical key: Symmetric-key algorithms are algorithms for cryptography that use the same cryptographic keys for both encryption of plaintext and decryption of ciphertext.

TCP/IP: Transmission control protocol/IP, used to govern the assembly of computer systems to the Internet.

Telemetry Transport or Message Queuing Telemetry Transport (MQTT/MQ): MQTT is an open OASIS and ISO standard (ISO/IEC 20922) lightweight, publish-subscribe network protocol that transports messages among devices.

Temporal Key Integrity Protocol (TKIP): TKIP is a security protocol used in the IEEE 802.11 wireless networking standard.

Token-Based Passwords (OTP): OTP is a piece of hardware called a security token. Inside the token is an accurate clock that has been synchronized with the clock on the proprietary verification server.

Transmission Control Protocol (TCP): TCP is one of the main protocols of the IP suite. It originated in the initial network implementation in which it complemented the IP. Therefore, the entire suite is commonly referred to as TCP/IP.

Transport Layer Security (TLS): TLS and Secure Sockets Layer (SSL are cryptographic protocols designed to provide communications security over a computer network).

Triple DES (3DES or TDES): Triple DES (3DES or TDES) is a symmetric-key block cipher. It applies the DES cipher algorithm three times to each data block.

Trojans: Trojan is a malware that misleads users of its true intent.

Two-Factor Authentication (2FA): Tw2FA is a multi-factor authentication (MFA) that reinforces access security by requiring two methods to verify user identity.

Twofish (Cipher): Twofish is a symmetric key block cipher with a block size of 128 bits and key sizes up to 256 bits.

Uniform Resource Locator (URL): A URL is a standardized naming convention for addressing documents available over the Internet and Intranet.

User Datagram Protocol (UDP): UDP is a communications protocol that is primarily used for establishing low-latency and loss-tolerating connections between applications on the Internet. It speeds up transmissions by enabling the allocation of data before an agreement is provided by the receiving party.

Virtual Private Network (VPN): VPN is a private network across a public network and allows users to send and receive data.

Vishing: Vishing (voice or VoIP phishing) is like phishing and is carried out using voice technology. A vishing attack can be showed by voice email, VoIP (voice over IP), or landline or cellular telephone.

WannaCry Attack: The WannaCry ransomware attack in 2017 targeted computers running the Microsoft Windows operating system by encrypting information and demanding ransom payments in the Bitcoin cryptocurrency.

WannaCry Ransomware: The WannaCry ransomware attack in 2017 targeted computers running the Microsoft Windows operating system by encrypting data and demanding ransom payments in the Bitcoin cryptocurrency.

Web API Security: Focuses on the transfer of data through APIs that are connected to the Internet.

Web Application Firewall (WAF): A WAF Application Firewall helps protect web applications by filtering and monitoring HTTP traffic between a web request and the Internet.

Web Crawlers: A web crawler is a package or automated script which browses the World Wide Web in a methodical, automated manner.

Web Crawling Phishing Attack Detector (WC-PAD): WC-PAD takes the web traffics, web content, and Uniform Resource Locator (URL) as input features, based on these features' classification of phishing and non-phishing websites are done.

Wi-Fi Protected Access (WPA): WPA is a security standard for users of computing devices armed with wireless Internet connections.

Wi-Fi Protected Access 2 (WPA2): WPA2 is the security method added to WPA for wireless networks that provide stronger data protection and network access control.

Wi-Fi Protected Access 3 (WPA3): WPA3 are three security and security certification programs developed by the Wi-Fi Alliance to secure wireless computer networks.

Wired Equivalent Privacy (WEP): WEP is a security algorithm for IEEE 802.11 wireless networks.

Wireless Intrusion Prevention Systems (WIPS): WIPS is a network device that monitors the radio spectrum for the presence of unauthorized access points and can automatically take countermeasures.

Wireless Sensor Network (WSN): WSN refers to a group of spatially dispersed and dedicated sensors. There are numerous wireless standards and solutions for sensor node connectivity.

Worms: A computer worm is a standalone malware program that replicates itself in order to spread to other computers.

Yubikeys: USB device that generates a unique passcode.

Zukey: It is a security token hardware.

NAMES

Sergey Aleynikov: Sergey Aleynikov is a former Goldman Sachs computer programmer. Between 2009 and 2016, he was twice impeached for the same conduct of allegedly copying proprietary computer source code from his employer, Goldman Sachs, before joining a competing firm.

James P. Anderson: J. P. Anderson outlined a series of definitive steps in computer security technology planning for US government.

Atari ST: The Atari ST is a line of home computers from Atari Corporation. The Atari ST is part of a mid-1980s generation of home computers.

Giovan Battista Bellaso: Giovan Battista Bellaso was an Italian cryptologist.

Alphonse Bertillon: Alphonse Bertillon was a French police officer and biometrics researcher who applied the anthropological method of anthropometry to law enforcement creating an identification system based on physical measurements.

Walter Bruch: He was a German electrical engineer the creator of Closed-circuit television. He invented the PAL color television system at Telefunken in the early 1960.

Fred Cohen: Frederick B. Cohen is an American computer scientist and finest known as the inventor of computer virus defense techniques.

Fernando Corbató: Fernando José "Corby" Corbató (July 1, 1926–July 12, 2019) was a prominent American computer scientist, notable as an innovator in the development of time-sharing operating systems.

Dorothy Denning: Dorothy Elizabeth Denning is a data security researcher known for lattice-based access control (LBAC), intrusion detection systems (IDS).

Glenn Greenwald: Glenn Edward Greenwald is an American journalist, author, and former attorney. He is best known for issuing a series of reports detailing previously unknown information about American and British global surveillance programs.

Edward Hugh Hebern: Edward Hugh Hebern was inventor of rotor machines, devices for encryption.

Edward Henry: Sir Edward Richard Henry was the Commissioner of Police of the Metropolis (head of the Metropolitan Police of London) from 1903 to 1918.

William Herschel: William Herschel, while working in colonial India, recognized the unique qualities that fingerprints had to proposal as a means of identification in the late 1870s. He first began using fingerprints as a form of signature on contracts with locals.

Kexue Huang: Kexue Huang worked at a Dow Chemical Co subsidiary from 2003 to 2008 in Indiana where he led a team of scientists developing biological insecticides and then later for another agribusiness giant, privately held Cargill Inc. He pleaded guilty in a federal court in Indiana to one count of stealing trade secrets from Cargill and one count of engaging in economic espionage at Dow.

Thomas Jefferson: Thomas Jefferson (April 13, 1743–July 4, 1826) was an American statesman, diplomat, lawyer, architect, philosopher, and Founding Father who served as the third president of the United States from 1801 to 1809. He formerly served as the second vice president of the United States from 1797 to 1801.

Ewen MacAskill: Ewen MacAskill a political editor who was involved in preparing the publication disclosures from Edward Snowden of the activities of the American National Security Agency (NSA).

Jeff Mogul: Jeff is the author or co-author of numerous Internet Standards; he contributed extensively to the HTTP/1.1 specification.

Robert Morris: Robert H. Morris Sr. was an American cryptographer and computer scientist.

Elon Musk: Elon Musk co-founded and leads Tesla, SpaceX, Neuralink, and The Boring Company.

Peter Neumann: Peter Gabriel Neumann is a computer-science researcher who worked on the Multics operating system in the 1960s.

Laura Poitras: Laura Poitras is a director and creator of documentary films. Laura Poitras is famous for Oscar-winning "Citizenfour," documentary on Edward Snowden story.

Dave Rand: David G. Rand is an associate professor of management science and brain and cognitive sciences at Massachusetts Institute of Technology.

Rivest–Shamir–Adleman: Ron Rivest, Adi Shamir, and Leonard Adleman are inventors of the RSA algorithm.

John Romkey: John Romkey, along with Donald W. Gillies, developed MIT PC/IP, the first TCP/IP stack in the industry for MS-DOS on the IBM PC[2][3][4][5] in 1983 while at the Massachusetts Institute of Technology.

Arthur Scherbius: Arthur Scherbius was a German electrical engineer who invented the mechanical cipher Enigma machine. He patented the invention and later sold the machine under the brand name Enigma.

Allan Scherr: Allan L. Scherr is an American computer scientist distinguished for his work in time-sharing operating systems and leading the original development of the IBM MVS operating system, used on IBM mainframe computers.

Bruce Schneier: Bruce Schneider is an American cryptographer, computer security professional, privacy specialist, and writer.

Edward Snowdown: Edward Joseph Snowden is an American whistleblower who copied and leaked highly classified information from the National Security

Agency (NSA) in 2013 when he was a Central Intelligence Agency (CIA) employee and subcontractor.

Ray Tomlinson: Raymond Samuel Tomlinson was an American computer programmer who implemented the first email program on the ARPANET system, the precursor to the Internet, in 1971; he is internationally known and credited as the inventor of email.

Alan Turing: Alan Mathison Turing was an English mathematician, computer scientist, logician, cryptanalyst, philosopher, and theoretical biologist.

Paul Vixie: Paul Vixie is an American computer scientist whose technical contributions include Domain Name System protocol design and procedure, mechanisms to attain operational robustness of DNS implementations, and significant contributions to open-source software principles and methodology.

Watergate: The Watergate scandal was a political scandal in the United States involving the administration of US President Richard Nixon from 1972 to 1974 that led to Nixon's resignation.

D.C. Williams: D.C. Williams was a stockbroker imprisoned for investigating on corporate telegraph lines and sold the information he overheard to stock traders.

Mike Yu: Xiang Dong Yu a former Ford employee who pleaded guilty to two counts of theft of trade secrets.

PRODUCT NAMES (PRODUCT NAMES CITED IN THIS BOOK)

Angry IP Scan (https://angryip.org/)
AppArmor (https://www.apparmor.com/)
Avast (https://www.avast.com/)
Barracuda Essentials (https://www.barracuda.com/)
Bitdefender (https://www.bitdefender.com/)
Censys (https://censys.io/)
Cisco ASA (https://www.cisco.com)
Cisco Cloud Email Security (https://www.cisco.com/)
Exabeam (https://www.exabeam.com/)
Fail2ban (https://www.fail2ban.org/)
Forcepoint Email Security (https://www.forcepoint.com/)
Google's Safe (https://safebrowsing.google.com/)
Malwarebytes (https://www.malwarebytes.com/)
McAfee (https://www.mcafee.com/)
Netcat (https://nmap.org/ncat/)
Network Mapper(https://nmap.org/)
OpenBSD (https://www.openbsd.org/)
pfSense (https://www.pfsense.org/)
PhishNet (https://phish.net/)
Proofpoint Essentials (https://www.proofpoint.com)
ScanGuard (https://www.scanguard.com/)
SELinux (https://www.redhat.com/)

Shodan (https://www.shodan.io/)
SonicWall (https://www.sonicwall.com/)
Sophos (https://home.sophos.com/)
SplashID (https://splashid.com/)
Splunk (https://www.splunk.com/)
StopBadware (/www.stopbadware.org/)
Thingful (https://www.thingful.net/)
TOTAL AV (https://www.totalav.com/)
Unicornscan (http://www.unicornscan.org/)
Zenmap (https://nmap.org/zenmap/)
ZoneAlarm (https://www.zonealarm.com/)
Zoomeye (https://www.zoomeye.org/)

References

CHAPTER 1: CYBERSECURITY TECHNOLOGIES CLASSIFICATION

1. Suryotrisongko H. and Musashi Y. (2014): "Review of Cybersecurity Research Topics, Taxonomy and Challenges: Interdisciplinary Perspective," 2019 IEEE 12th Conference on Service-Oriented Computing and Applications (SOCA), Kaohsiung, Taiwan, 2019, pp. 162–167. DOI: 10.1109/SOCA.2019.00031. https://ieeexplore.ieee.org/abstract/document/8953034.
2. Jouinia M., Ben Rabaia L. and Ben Aissab A. (2014): "Classification of Security Threats in Information Systems," Procedia Computer Science, Volume 32, pp. 489–496. https://www.sciencedirect.com/science/article/pii/S1877050914006528?via%3Dihub.

CHAPTER 2: ENCRYPTION

1. Whittaker Z. (2019): "Facebook Admits It Stored 'Hundreds of Millions of Account Passwords in Plaintext," Techcrunch.com, March 21, 2019. https://techcrunch.com/2019/03/21/facebook-plaintext-passwords/#:~:text=Facebook%20confirmed%20Thursday%20in%20a,of%20a%20routine%20security%20review.
2. Turing D. (2018): "X, Y & Z: The Real Story of How Enigma Was Broken," The History Press, Great Britain.
3. Daemen J. and Rijmen V. (2013): "The Design of Rijndael: AES – the Advanced Encryption Standard," Springer Science & Business Media, New York, NY.
4. IBM® Knowledge Center (2018): "About Encryption Keys," IBM 3592 E07/EH7, E08/EH8, 55F, and 60F Tape Drives documentation, December 7, 2018. https://www.ibm.com/support/knowledgecenter/en/STPRH6/com.ibm.storage.drives.doc/top_tscom_reuse_encryptoview_keys.html.
5. Singh G. and Supriya (2013): "A Study of Encryption Algorithms (RSA, DES, 3DES and AES) for Information Security," International Journal of Computer (0975 – 8887), Volume 67, No. 19, pp 32–38. https://www.semanticscholar.org/paper/A-Study-of-Encryption-Algorithms-(RSA%2C-DES%2C-3DES-Singh-Supriya/187d26258dc57d794ce4badb094e64cf8d3f7d88. DOI: 10.5120/11507-7224Corpus ID: 7625130.
6. Abdul Wahid M. N. et al. (2018): "A Comparison of Cryptographic Algorithms: DES, 3DES, AES, RSA and Blowfish for Guessing Attacks Prevention," Journal of Computer Science Applications and Information Technology. https://symbiosisonlinepublishing.com/computer-science-technology/computerscience-information-technology32.php.
7. Geeksforgeeks: "Blowfish Algorithm with Examples," geeksforgeeks.org. https://www.geeksforgeeks.org/blowfish-algorithm-with-examples/.
8. Zhou X. and Tang X. (2011, August): "Research and Implementation of RSA Algorithm for Encryption and Decryption." In Strategic Technology (IFOST), 2011 6th International Forum on IEEE, Volume 2, pp. 1118–1121.
9. Hern A. (2019): "Facebook Stored Hundreds of Millions of Passwords Unprotected," The Guardian, March 21, 2019. https://www.theguardian.com/technology/2019/mar/21/facebook-admits-passwords-unprotected.
10. Gartenberg A. (2018): "Twitter Advising All 330 Million Users to Change Passwords After Bug Exposed Them in Plain Text," The Verge, May 3, 2018. https://www.theverge.com/2018/5/3/17316684/twitter-password-bug-security-flaw-exposed-change-now.

11. Al-Heeti A. and Ng A. (2018): "Twitter Password Bug Potentially Exposes 330M Users, Jack Dorsey says," *CNET*, May 4, 2018. https://www.cnet.com/news/twitter-advises-all-users-to-change-passwords-after-glitch-that-exposed-some-in-plain-text.

CHAPTER 3: AUTHENTICATION

1. Nachreiner C. (2018): "Digital Authentication: The Past, Present and Uncertain Future of The Keys to Online Identity", geekwire.com, September 22, 2018. https://www.geekwire.com/2018/digital-authentication-human-beings-history-trust/.
2. Zekri L. and Furnell S. (2006): "Authentication Based Upon Secret Knowledge and Its Resilience to Impostors", Advances in Network & Communication Engineering, Volume 3, pp. 30–38.
3. Nelson D., Reed V. and Walling J. (1976): "Pictorial Superiority Effect," Journal of Experimental Psychology Human Learning and Memory, Volume 2, No. 5, pp. 523–528.
4. Aloul F., Zahidi S. and El-Hajj W. (2009): "Two Factor Authentication Using Mobile Phones," 2009 IEEE/ACS International Conference on Computer Systems and Applications, 2009, pp. 641–644.
5. "Secure Key: Two-Factor Authentication HSBC UK", HSBC Bank plc, [online] Available: https://www.hsbc.co.uk/help/security-centre/secure-key.
6. Harwell D. (2019): "Federal Study Confirms Racial Bias Of Many Facial-Recognition Systems, Casts Doubt on Their Expanding Use," Washington Post, December 19, 2019. https://www.washingtonpost.com/technology/2019/12/19/federal-study-confirms-racial-bias-many-facial-recognition-systems-casts-doubt-their-expanding-use/.
7. Bhattacharyya D., Ranjan R., Das P., Kim T. and Bandyopadhyay S. K. (2009): Biometric Authentication Techniques and its Future Possibilities," 2009 Second International Conference on Computer and Electrical Engineering, Dubai, 2009, pp. 652–655.
8. Bachmann M. (2014): "Passwords are Dead: Alternative Authentication Methods," 2014 IEEE Joint Intelligence and Security Informatics Conference, The Hague, 2014, pp. 322–322.
9. Bandyopadhyay S. K., Bhattacharyya D. and Das P. (2008): "User Authentication by Secured Graphical Password Implementation," 2008 7th Asia-Pacific Symposium on Information and Telecommunication Technologies, Bandos Island, 2008, pp. 7–12.
10. Ibrokhimov S., Hui K. L., Abdulhakim Al-Absi A., lee h. j. and Sain M. (2019): "Multi-Factor Authentication in Cyber Physical System: A State of Art Survey," 2019 21st International Conference on Advanced Communication Technology (ICACT), Pyeong Chang.
11. McLean R. (2017): "A Hacker Gained Access to 100 Million Capital One Credit Card Applications and Accounts," JCNN Business, July 30, 2019. https://edition.cnn.com/2019/07/29/business/capital-one-data-breach/index.html.
12. capitalone.com (2019): "Information on the Capital One Cyber Incident," capitalone.com, September 23, 2019. https://www.capitalone.com/facts2019/.

CHAPTER 4: BIOMETRICS

1. Dargan S. and Kumar M. (2020): "A Comprehensive Survey on the Biometric Recognition Systems Based on Physiological and Behavioral Modalities," Expert Systems with Applications, 2020, Volume 143, pp. 113–114. https://www-sciencedirect-com.libaccess.sjlibrary.org/science/article/pii/S0957417419308310.
2. Bhattacharyya D., Ranjan R., Das P., Kim T. H. and Bandyopadhyay S. K. (2009): "Biometric Authentication Techniques and Its Future Possibilities," 2009 Second International Conference on Computer and Electrical Engineering, 2009. https://ieeexplore-ieee-org.libaccess.sjlibrary.org/document/5380550.

3. Wilkinson C. J. (2018): "Airport Staff Access Control: Biometrics at Last?" 2018 International Carnahan Conference on Security Technology (ICCST), Montreal, QC, 2018, pp. 1–8. https://ieeexplore-ieee-org.libaccess.sjlibrary.org/document/8585592.
4. Nainan S., Ramesh A., Gohil V. and Chaudhary J. (2017): "Speech Controlled Automobile with Three-Level Biometric Security System," 2017 International Conference on Computing, Communication, Control and Automation (ICCUBEA), Pune, 2017, pp. 1–6. https://ieeexplore-ieee-org.libaccess.sjlibrary.org/document/8464003.
5. Robertson J. J., Guest R. M., Elliott S. J. and O'Connor K. (2017): "A Framework for Biometric and Interaction Performance Assessment of Automated Border Control Processes," IEEE Transactions on Human-Machine Systems, December 2017, Volume 47, No. 6, pp. 983–993. https://ieeexplore-ieee-org.libaccess.sjlibrary.org/document/7592857.
6. Hill K. (2020): "Wrongfully Accused by an Algorithm," New York Times, August 3, 2020. https://www.nytimes.com/2020/06/24/technology/facial-recognition-arrest.html.

CHAPTER 5: FIREWALL TECHNOLOGIES

1. Clement J. (2020): "Internet Usage Worldwide – Statistics & Facts," statista.com. https://www.statista.com/topics/1145/internet-usage-worldwide/.
2. Tharaka S. C. et al. (2016): "High-Security Firewall: Prevent Unauthorized Access Using Firewall Technologies," International Journal of Scientific and Research Publications, Volume 6, No. 4, 504. ISSN 2250-3153. http://www.ijsrp.org/research-paper-0416.php?rp=P525293.
3. Lee H. K. et al. (2007): "Cryptographic Strength of SSL/TLS Servers: Current and Recent Practices," IMC '07: Proceedings of the 7th ACM SIGCOMM Conference on Internet Measurement, October 2007, pp. 83–92. https://doi.org/10.1145/1298306.1298318.
4. Grigorik I. (2013): "Performance Browser Networking," O'Reilly, 2013. https://hpbn.co/transport-layer-security-tls/.
5. Arsene L. (2015): "The Evolution of Firewalls: Past, Present & Future," Information Week, January 27, 2015. https://www.informationweek.com/partner-perspectives/the-evolution-of-firewalls-past-present-and-future/a/d-id/1318814.
6. Kovacs Ed. (2019): "Cyberattack Disrupted Firewalls at U.S. Power Utility, Security Week," September 9, 2019. https://www.securityweek.com/cyberattack-disrupted-firewalls-us-power-utility.
7. Kovacs Ed. (2019): "Cisco Firewall Exploited in Attack on U.S. Renewable Energy Firm," Security Week, November 1, 2019. https://www.securityweek.com/cisco-firewall-vulnerability-exploited-attack-us-renewable-energy-provider.
8. Cimpanu C. (2019): "Cyber-Security Incident at US Power Grid Entity Linked to Unpatched Firewalls," ZDNET, September 9, 2019.

CHAPTER 6: VIRUS DETECTION

1. Harrison V. and Jose Pagliery J. (2015): "Nearly 1 Million New Malware Threats Released Every Day," CNN Business, April 14, 2015. https://money.cnn.com/2015/04/14/technology/security/cyber-attack-hacks-security/.
2. Metcalf John (2014): "Core War – Creeper and Reaper," Corewar.co.uk. https://corewar.co.uk/creeper.htm.
3. Cohen F. (1987): "Computer Viruses Theory and Experiments," Computers and Security, Volume 6, pp. 22–35. https://dl.acm.org/doi/10.1016/0167-4048%2887%2990122-2.
4. "Detecting Computer Viruses," International Journal of Advanced Research in Computer Engineering & Technology (IJARCET), July 2014, Volume 3, No. 7. http://ijarcet.org/wp-content/uploads/IJARCET-VOL-3-ISSUE-7-2357-2364.pdf.

5. Zetter K. (2014): "An Unprecedented Look at Stuxnet, the World's First Digital Weapon," WIRED, November 3, 2014.

CHAPTER 7: PHISHING DETECTION

1. Anti-Phishing Working Group (APWG) (2020): "Phishing Activity Trends Report – Fourth Quarter 2019, Activity October–December 2019, February 24, 2020. https://docs.apwg.org/reports/apwg_trends_report_q4_2019.pdf.
2. Jain A. and Gupta B. (2017): "Phishing Detection: Analysis of Visual Similarity Based Approaches. Security and Communication Networks," 2017, pp. 1–20. https://www.hindawi.com/journals/scn/2017/5421046/.
3. Stutz M. (1998): "AOL: A Cracker's Paradise?" Wired News, January 29, 1998. http://wired-vig.wired.com/news/technology/0,1282,9932,00.html.
4. Zetter T. (2009): "Bullion and Bandits: The Improbable Rise and Fall of E-Gold," WIRED, June 9, 2019. https://www.wired.com/2009/06/e-gold/.
5. Prakash P., Kumar M., Kompella R. R. and Gupta M. (2010): "PhishNet: Predictive Blacklisting to Detect Phishing Attacks," 2010 Proceedings IEEE INFOCOM, San Diego, CA, pp. 1–5. DOI: 10.1109/INFCOM.2010.5462216.
6. Felegyhazi M., Kreibich C. and Paxson V. (2010): "On the Potential of Proactive Domain Blacklisting," International Computer Science Institute. https://www.icsi.berkeley.edu/icsi/node/4554.
7. Mirtaheri S. M., Dincturk M. E., et al. (2013): "A Brief History of Web Crawlers," Proceedings of CASCON 2013, Toronto, November 2013.
8. Nathezhtha T., Sangeetha D. and Vaidehi V. (2019): "WC-PAD: Web Crawling Based Phishing Attack Detection," 2019 International Carnahan Conference on Security Technology (ICCST), Chennai, India, pp. 1–6. DOI: 10.1109/CCST.2019.8888416.
9. Huddleston T. (2019): "Sorry I Missed That Here We Are How This Scammer Used Phishing Emails to Steal Over $100 Million from Google and Facebook," March 27, 2019. https://www.cnbc.com/2019/03/27/phishing-email-scam-stole-100-million-from-facebook-and-google.html.

CHAPTER 8: ENDPOINT PROTECTION

1. Sharron Malaver S. (2018): "A Historical Take on the Evolution of Endpoint Security," minerva-labs.com, July 5, 2018. https://blog.minerva-labs.com/a-historical-take-on-the-evolution-of-endpoint-security.
2. Barros A. and Chuvakin A. (2016): "Comparison of Endpoint Detection and Response Technologies and Solutions," Gartner, June 10, 2016. https://www.gartner.com/en/documents/3343417.
3. Soffar H. (2019): "The Advantages and Disadvantages of Anti-virus software," Online-Science, November 6, 2019. https://www.online-sciences.com/computer/the-advantages-and-disadvantages-of-anti-virus-software/.
4. Litan A. (2017): "Booming $500 Million EDR Market Faces Stiff Challenges," Gartner, January 12, 2017. https://blogs.gartner.com/avivah-litan/2017/01/12/booming-500-million-edr-market-faces-stiff-challenges/.
5. Taylor H. (2029): "IDC says 70% of Successful Data Breaches Occur Through Endpoints," September 9, 2019 Off Cyber Security Research," Journal of Cyber Policy, September 9, 2019. https://journalofcyberpolicy.com/2019/09/09/idc-says-70-successful-breaches-originate-endpoint/.
6. Specter M. A. and Halderman J. Alex. (2020): "Security Analysis of the Democracy Live Online Voting System," Internetpolicy.mit.edu, June 6, 2020. https://internetpolicy.mit.edu/wp-content/uploads/2020/06/OmniBallot.pdf.

7. Hautala L. (2020): "Bad Security Design Mars OmniBallot Online Voting System, Report Says," CNET, June 8, 2010. https://www.cnet.com/news/critical-security-flaws-mar-omniballot-online-voting-system-report-says/.

CHAPTER 9: MALWARE PROTECTION TECHNOLOGY

1. MacAfee: "What Is the Difference Between Malware and a Virus?" www.mcafee.com. https://www.mcafee.com/enterprise/en-us/security-awareness/ransomware/malware-vs-viruses.html#:~:text=viruses%20is%20an%20important%20one,its%20code%20into%20other%20programs.
2. Love J. (2018): "A Brief History of Malware — Its Evolution and Impact," lastline.com, April 5, 2018. https://www.lastline.com/blog/history-of-malware-its-evolution-and-impact/.
3. "Anti-Virus Is 30 Years Old," spgedwards.com, April, 2012. Archived from the original on April 27, 2015. [s5] "A Brief History of Antivirus Software". techlineinfo.com. Archived from the original on August 26, 2014.
4. Pirc J. (2017): "The Evolution of Intrusion Detection/Prevention: Then, Now and the Future," Secureworks, July 6, 2017. Retrieved from https://www.secureworks.com/blog/the-evolution-of-intrusion-detection-prevention.
5. Shevchenko A. (2007): "The Evolution of Technologies Used to Detect Malicious Code," securelist.com, Kaspersky Lab, November 7, 2007. https://securelist.com/the-evolution-of- technologies-used-to-detect-malicious-code/36177/.
6. Greenberg A. (2017): "'Crash Override': The Malware That Took Down a Power Grid," WIRED, June 12, 2017. https://www.wired.com/story/crash-override-malware/.
7. FIREEYE (2020): "Highly Evasive Attacker Leverages SolarWinds Supply Chain to Compromise Multiple Global Victims with SUNBURST Backdoor," Fireeye.com, December 13, 2020. https://www.fireeye.com/blog/threat-res.
8. Satter R., Bing C. H. and Menn J. (2020): "Hackers Used SolarWinds' Dominance Against It in Sprawling Spy Campaign." Reuters, December 15, 2020. https://www.reuters.com/article/idUSKBN28Q07P.

CHAPTER 10: THE INTERNET OF THINGS

1. Gartner (2017): "Gartner Says 8.4 Billion Connected "Things" Will Be in Use in 2017, Up 31 Percent from 2016," Gartner, Egham, UK. https://www.gartner.com/en/newsroom/press-releases/2017-02-07-gartner-says-8-billion-connected-things-will-be-in-use-in-2017-up-31-percent-from-2016.
2. "The Internet of Things (IoT): An Overview," Cong. Research Serv., February 2020. https://crsreports.congress.gov/product/pdf/IF/IF11239.
3. Schiffer A. (2017): "How a Fish Tank Helped Hack a Casino," Washington Post, July 2017. https://www.washingtonpost.com/news/innovations/wp/2017/07/21/how-a-fish-tank-helped-hack-a-casino/.
4. Simon S. (2016): "'Internet of Things' Hacking Attack Led to Widespread Outage of Popular Websites," Weekend Edition, NPR, Saturday, October 2016. https://www.npr.org/2016/10/22/498954197/internet-outage-update-internet-of-things-hacking-attack-led-to-outage-of-popula.
5. Moyer C. (2017): "This Teen Hacked 150,000 Printers to Show How the Internet of Things Is Shit," VICE, February, 2017. https://www.vice.com/en_us/article/nzqayz/this-teen-hacked-150000-printers-to-show-how-the-internet-of-things-is-shit.
6. Harris K. D. (2016): "California Data Breach Report," State of California – Department of Justice, February 2016 "In 2014, 67 percent of breach victims in the U.S. were also victims of fraud, compared to just 25 percent of all consumers."

7. Ethan Castro E. (2017): "Hidden History: John Romkey and the Internet Toaster," TechNews, November 8, 2017. https://www.technewsiit.com/node/1527.
8. Pres G. (2014): "A Very Short History of the Internet of Things," Forbes, June 18, 2014. https://www.forbes.com/sites/gilpress/2014/06/18/a-very-short-history-of-the-internet-of-things/#7c9ef45010de.
9. Vuppalapati Ch. (2019): "Building Enterprise IoT Applications," Taylor & Francis, 2019, ISBN: 9780429508691.
10. Sicari S., Rizzardi A., Grieco L. and Coen-Porisini A. (2015): "Security, Privacy and Trust in Internet of Things: The Road Ahead," Computer Networks, Volume 76, pp. 146–164.
11. Wójtowicz A. and Cellary W. (2019): "New Challenges for User Privacy in Cyberspace," Human Computer Interaction and Cybersecurity Handbook, edited by Abbas Moallem, 1st ed., CRC Press, New York, NY, pp. 79–80.
12. Schiffer A. (2017): "How a Fish Tank Helped Hack a Casino," Washington Post, July 21, 2017. https://www.washingtonpost.com/news/innovations/wp/2017/07/21/how-a-fish-tank-helped-hack-a-casino/?noredirect=on.
13. Stanley A. (2020): "This Hacked Coffee Maker Demands Ransom and Demonstrates a Terrifying Implication About the IoT," GIZMODO, September 26, 2020.

CHAPTER 11: NETWORK SECURITY

1. Wikipedia (2020): "List of Security Hacking Incidents," March 14, 2020. https://en.wikipedia.org/wiki/List_of_security_hacking_incidents.
2. Afshar Alam M., Siddiqui T. and Seeja K. R. (2009): "Recent Developments in Computing and Its Applications," I. K. International Pvt Ltd., pp. 513. ISBN 978-93-80026-78-7.
3. CBS (2020): "San Jose Programmer Pleads Guilty to Damaging Cisco's Network," San Francisco, CBS Local, August 26, 2020. https://sanfrancisco.cbslocal.com/2020/08/26/san-jose-programmer-pleads-guilty-to-damaging-ciscos-network/.

CHAPTER 12: LOCATION TRACKING

1. Bajaj R., Ranaweera S. L. and Agrawal D. P. (2002): "GPS: Location-tracking Technology," Computer, Volume 35, No. 4, pp. 92–94. https://ieeexplore.ieee.org/abstract/document/993780.
2. Bruce Morser B. (2020): "How Does GPS Work?" Smithsonian Institution, March 2020. https://timeandnavigation.si.edu/multimedia-asset/how-does-gps-work.
3. Kharpal A. (2020): "China Aims to Complete Its Own GPS System, Giving Beijing Military Independence in Case of Conflict," CNBC, June 21, 2020. https://www.cnbc.com/2020/06/22/beidou-china-aims-to-complete-gps-system-that-rivals-us.html.
4. Kim S., Jeong Y. and Park S. (2020): "RFID-Based Indoor Location Tracking to Ensure the Safety of the Elderly in Smart Home Environments," Personal and Ubiquitous Computing, Volume 17, March 2020, pp. 1699–1707. https://doi.org/10.1007/s00779-012-0604-4.
5. Dhar S. and Upkar V. (2011): "Challenges and Business Models for Mobile Location-Based Services and Advertising," Communications of the ACM, Volume 54, No. 5, pp. 121–128. DOI: 10.1145/1941487.
6. Leung H. (2019): "What to Know About Absher, Saudi Arabia's Controversial 'Woman-Tracking' App," Time, February 19, 2019. https://time.com/5532221/absher-saudi-arabia-what-to-know/.
7. Bennett C. (2019): "Wife-Tracking Apps are One Sign of Saudi Arabia's Vile Regime. Others Include Crucifixion …," The Guardian, April 28, 2019. https://www.theguardian.com/commentisfree/2019/apr/28/wife-tracking-apps-saudi-arabias-vile-regime-crucifixion.

8. Tau B. (2020): "Government Tracking How People Move Around in Coronavirus Pandemic," Wall Street Journal, March 28, 2020. https://www.wsj.com/articles/government-tracking-how-people-move-around-in-coronavirus-pandemic-11585393202.

CHAPTER 13: SURVEILLANCE

1. Zuboff S. (2019): "The Age of Surveillance Capitalism, The Fight for a Human Future at the New Frontier of Power," published by Profile, 2019, pp. 7.
2. BECKY Little B. (2018): "Communications Companies Have Been Spying on You Since the 19th Century," History.com, August 30, 2018. https://www.history.com/news/communications-companies-have-been-spying-on-you-since-the-19th-century#:~:text=But%20they%20also%20intercepted%20morse,meant%20to%20deceive%20the%20enemy.&text=In%201864%2C%20a%20stockbroker%20named,sold%20them%20to%20stock%20traders.
3. Auerbach J. S. (1964): "The La Follette Committee: Labor and Civil Liberties in the New Deal," The Journal of American History, Volume 51, No. 3 (December 1964), pp. 435–459 (25 pages), Oxford University Press on behalf of Organization of American Historians.
4. Young J. (2018): "A History of CCTV Surveillance in Britain," SWNS, January 22, 2018. https://stories.swns.com/news/history-cctv-surveillance-britain-93449/.
5. Bradford L. (2019): "A History of CCTV Technology: How Video Surveillance Technology Has Evolved." Surveillance, 5 December 2019. www.surveillance-video.com/blog/a-history-of-cctv-technology-how-video-surveillance-technology-has-evolved.html/.
6. Theoharis A. G.: "Abuse of Power: How Cold War Surveillance and Secrecy Policy Shaped the Response to 9/11," Temple University Press. pp. 212.
7. Zetter K. (2016): "Everything We Know About How the FBI Hacks People," WIRED, May 2016. https://www.wired.com/2016/05/history-fbis-hacking/.
8. Ambinder M. (2009): "Pinwale and The New NSA Revelations," The Atlantic, June 16, 2009. https://www.theatlantic.com/politics/archive/2009/06/pinwale-and-the-new-nsa-revelations/19532/.
9. Poulsen K. (2007): "FBI's Magic Lantern Revealed," WIRED, July 2007. https://www.wired.com/2007/07/fbis-magic-lant/.
10. Gellman B. and Poitras June L. (2013): "U.S., British Intelligence Mining data from nine U.S. Internet Companies in Broad Secret Program," Washington Post, June 7, 2013. https://www.washingtonpost.com/investigations/us-intelligence-mining-data-from-nine-us-internet-companies-in-broad-secret-program/2013/06/06/3a0c0da8-cebf-11e2-8845-d970ccb04497_story.html.
11. Adida B., Anderson C., Anton A.I., Blaze M. and Dingledine R. (2013): "CALEA II: Risks of Wiretap Modifications to Endpoints," cyberwar.nl, May 17, 2013. https://cyberwar.nl/d/calea2-wiretap.pdf.
12. O'Brien L. (2007): "FBI Confirms Contracts with AT&T, Verizon and MCI," WIRED, March 20, 2007. https://www.wired.com/2007/03/fbi-confirms-co/.
13. Lewis R. (2015): "Stingray Cell Surveillance Can Record Conversations, Bug Phones," Aljazeera America, October 29, 2015. http://america.aljazeera.com/articles/2015/10/29/stingray-can-spy-on-calls-and-texts-bug-phones.html#:~:text=Stingrays%2C%20or%20cell%20site%20simulators,recently%20obtained%20by%20the%20ACLU.
14. Lye L. (2015): "New Docs: DOJ Admits that StingRays Spy on Innocent Bystanders," ACLUNC, October 28, 2015. https://www.aclunc.org/blog/new-docs-doj-admits-stingrays-spy-innocent-bystanders.
15. Michel A. H. (2019): "Eyes in the Sky: The Secret Rise of Gorgon Stare and How It Will Watch Us All," Houghton Mifflin Harcourt, 2019.

16. Bogel-Burroughs N. (2020): "Baltimore Hopes Surveillance Planes Lower Crime, but Residents Fear Abuse," New York Times, April 9, 2020. https://www.nytimes.com/2020/04/09/us/baltimore-surveillance-planes-aclu.html.

17. Laperruque J. and Janovsky D. (2018): "These Police Drones Are Watching You," POGO, September 25, 2018. https://www.pogo.org/analysis/2018/09/these-police-drones-are-watching-you/.

18. Perrin C. (2008): "Wim van Eck's legacy," TechRepublic, October 23, 2008. https://www.techrepublic.com/blog/it-security/wim-van-ecks-legacy/.

19. Brandon R. (2018): "The Importance of Cybersecurity in Modern Video Surveillance Environments," Security Magazine RSS, Security Magazine, September 25, 2018. www.securitymagazine.com/articles/89444-the-importance-of-cybersecurity-in-modern-video-surveillance-environments.

20. ACLU (2002): "What's Wrong with Public Video Surveillance?," aclu.org, March, 2002. https://www.aclu.org/other/whats-wrong-public-video-surveillance.

21. Thomas-Lester A. and Locy T. (1997): "Chief's Friend Accused of Extortion," Washington post, November 26, 1997. https://www.washingtonpost.com/wp-srv/local/longterm/library/dc/dcpolice/stories/stowe25.htm.

22. Volokh E. (2002): "The Benefits of Surveillance," The Responsive Community, Fall 2002, pp. 9. https://www2.law.ucla.edu/volokh/camerascomm.htm.

23. Frances A. (2016): "Is There Empirical Evidence That Surveillance Cameras Reduce Crime?," Municipal Technical Advisory Service (MTAS), September 26, 2016. https://www.mtas.tennessee.edu/knowledgebase/there-empirical-evidence-surveillance-cameras-reduce-crime.

24. Vigdor N. (2019): "Somebody's Watching: Hackers Breach Ring Home Security Cameras," New York Times, December 15, 2019. https://www.nytimes.com/2019/12/15/us/Hacked-ring-home-security-cameras.html.

25. Cimpanue C. (2021): "Rogue CCTV Technician Spied on Hundreds of Customers During Intimate Moments," ZDNet, January 23, 2021. https://www.zdnet.com/article/rogue-cctv-technician-spied-on-hundreds-of-customers-during-intimate-moments/.

CHAPTER 14: INSIDER THREAT PROTECTION

1. Papadaki M. and Stavros Shiaeles S. (2018): "Insider Threat, The Forgotten, Yet Formidable Foe," In Human-Computer Interaction and Cybersecurity Handbook," CRC Press, 2018, pp. 124.

2. Klayman B. (2011): "Ex-Ford Engineer Sentenced for Trade Secrets Theft," Reuters, April 13, 2011. https://www.reuters.com/article/us-djc-ford-tradesecrets-idUSTRE73C3FG20110413.

3. Pelofsky J. (2011): "Chinese Man Pleads Guilty for US Trade Secret Theft. (2011b)," Reuters, October 18, 2011. https://www.reuters.com/article/us-crime-china-theft/chinese-man-pleads-guilty-for-u-s-trade-secret-theft-idUSTRE79H78R20111018.

4. Tsukayama H. (2012): "Chinese Hackers Breach Nortel Networks," Washington Post, February 14, 2012. https://www.washingtonpost.com/business/technology/report-chinese-hackers-breach-nortel-networks/2012/02/14/gIQApXsRDR_story.html.

5. Larry Greenemeier L. (2007): "Massive Insider Breach at DuPont," CRN, February 15, 2007. https://www.crn.com/news/security/197006655/massive-insider-breach-at-dupont.htm.

6. Greenwald G., MacAskill E. and Poitras L. (2013): "Edward Snowden: The Whistleblower Behind the NSA Surveillance Revelations," The Guardian, June 2013. https://www.theguardian.com/world/2013/jun/09/edward-snowden-nsa-whistleblower-surveillance

7. Zetter K. (2015): "Programmer Convicted in Bizarre Goldman Sachs Case – Again," WIRED, May 1, 2015. https://www.wired.com/2015/05/programmer-convicted-bizarre-goldman-sachs-caseagain/.

8. Brook C. (2020): "Tesla Data Theft Case Illustrates the Danger of the Insider Threat," Digital Guardian, January 27, 2020. https://digitalguardian.com/blog/tesla-data-theft-case-illustrates-danger-insider-threat.

9. Kolodny L. (2018): "Elon Musk Emails Employees About 'Extensive And Damaging Sabotage' by Employee," CNBC, June 18, 2018. https://www.cnbc.com/2018/06/18/elon-musk-email-employee-conducted-extensive-and-damaging-sabotage.html.

10. Huang J-W et al. (2018): "Email Security Level Classification of Imbalanced Data Using Artificial Neural Network: The Real Case in a World-Leading Enterprise," Engineering Applications of Artificial Intelligence, Volume 75, October 2018, pp. 11–12.

11. Greenberg A. (2020): "Tesla Employee Thwarted an Alleged Ransomware Plot," WIRED, August 27, 2020. https://www.wired.com/story/tesla-ransomware-insider-hack-attempt/

CHAPTER 15: INTRUSION DETECTION

1. Bruneau G. (2020): "The History and Evolution of Intrusion Detection," SANS Institute Information Security Reading Room, 2020. https://www.sans.org/reading-room/whitepapers/detection/history-evolution-intrusion-detection-344.

2. Ellis J. (2020): "State of the Practice of Intrusion Detection Technologies," January 2000 Networked Systems Survivability Program, January 25, 2000. https://resources.sei.cmu.edu/asset_files/TechnicalReport/2000_005_001_16796.pdf.

3. Boutaba R. and Fung C. (2013): "Intrusion Detection Networks," Auerbach Publications, 2013.

4. Skrobanek P. (2011): "Intrusion Detection Systems," Intech Open Publisher, March 22, 2011. ISBN: 978-953-307-167-1. https://www.intechopen.com/books/intrusion-detection-systems.

5. Harale N. and Meshram B.B. (2016): "Network Based Intrusion Detection and Prevention Systems: Attack Classification, Methodologies and Tools," Semantic Scholar, 2016. Corpus ID: 212537112. https://www.semanticscholar.org/paper/Network-Based-Intrusion-Detection-and-Prevention-%2C-Harale-Meshram/e87432283e90da5bceb3b17550570ec3a210f43c.

6. Johnson L. (2020): "Barnes & Noble Hit with Security Breach," RISNEWS, October 16, 2020. https://risnews.com/barnes-noble-hit-security-breach.

7. Hope A. (2020): "Barnes & Noble Alerted Customers of Data Breach That Leaked Personal and Transaction Information," October 20, 2020. https://www.cpomagazine.com/cyber-security/barnes-noble-alerted-customers-of-data-breach-that-leaked-personal-and-transaction-information/.

8. Osborne C. H. (2020): "Barnes & Noble Confirms Cyberattack, Ransomware Group Leaks Allegedly Stolen Data," ZDNET, October 20, 2020. https://www.zdnet.com/article/barnes-noble-confirms-cyberattack-customer-data-breach/.

CHAPTER 16: VULNERABILITY SCANNING

1. Bekerman D. and Yerushalmi S. (2019): "The State of Vulnerabilities in 2019," imperva.com, January 22, 2020. https://www.imperva.com/blog/the-state-of-vulnerabilities-in-2019/#:~:text=Figure%201%20shows%20the%20number,compared%20to%202017%20(14%2C086).

2. Horev R. (2019): "A History of Vulnerability Management," blog.vulcan.io, March 14, 2019. https://blog.vulcan.io/a-history-of-vulnerability-management#:~:text=In%20the%20late%2090s%20and,first%20vulnerability%20scanners%20were%20released.&text=For%20example%2C%20in%20the%20year,very%20much%20a%20manual%20process.

3. Shema M. (2014): "Anti-Hacker Tool Kit," McGraw-Hill, 4th ed., February 2014. ISBN: 9780071800150.
4. Tundis A. and Wojciech Mazurczyk W. (2018): "A Review of Network Vulnerabilities Scanning Tools: Types, Capabilities and Functioning," Proceedings of the 13th International Conference on Availability, Reliability and Security, August 2018, Article No.: 65, pp. 110. https://doi.org/10.1145/3230833.3233287.
5. Lis A. (2019): "Comparison and Analysis of Web Vulnerability Scanners," Institute of Information Security University of Stuttgart Universitätsstraße 38, D-70569 Stuttgart, Bachelorarbeit thesis November 20, 2019. https://elib.uni-stuttgart.de/handle/11682/10634.
6. DNS Stuff (2020): "Top 15 Paid and Free Vulnerability Scanner Tools in 2020," www.dnsstuff.com, January 6, 2020. https://www.dnsstuff.com/network-vulnerability-scanner.
7. OWASP: "Vulnerability Scanning Tools," owasp.org. https://owasp.org/www-community/Vulnerability_Scanning_Tools.
8. Kimball J. (2020): "8 of the Best Network Vulnerability Scanners Tested," comparitech.com. https://www.comparitech.com/net-admin/free-network-vulnerability-scanners/#Conclusion.
9. McLean R. (2019): "A Hacker Gained Access to 100 Million Capital One Credit Card Applications And Accounts," CNN Business, July 30, 2019. edition.cnn.com/2019/07/29/business/capital-one-data-breach/index.html.
10. Osborne C. H. (2019): "Severe Vulnerability in Apple FaceTime Found by Fortnite player the Teen's Mother Attempted to Contact Apple with No Success," ZDNET, January 30, 2019. https://www.zdnet.com/article/apple-facetime-exploit-found-by-14-year-old-playing-fortnite/.

CHAPTER 17: PENETRATION TESTING

1. Henry K. M. (2012): "Penetration Testing: Protecting Networks and Systems," IT Governance Publishing, 2012.
2. Anderson J. P. (1972): "Computer Security Technology Planning Study," October 1972, ESD-TR-7315 r, Volume II. https://csrc.nist.gov/csrc/media/publications/conference-paper/1998/10/08/proceedings-of-the-21st-nissc-1998/documents/early-cs-papers/ande72.pdf.
3. Zunnurain Hussain M. et al. (2017): "Penetration Testing in System Administration," International Journal of Scientific & Technology Research, Volume 6, No. 6. https://www.researchgate.net/profile/Muhammad_Zulkifl_Hasan2/publication/319876508_Penetration_Testing_In_System_Administration/links/59bf7cedaca272aff2e1b8de/Penetration-Testing-In-System-Administration.pdf.
4. Infosecinstitute. (2020): "14 Best Open-Source Web Application Vulnerability Scanners," Posted on July 13, 2020. https://resources.infosecinstitute.com/14-popular-web-application-vulnerability-scanners/#gref.
5. Weaver B. and Halton W. et al. (2017): "publisher logo Penetration Testing: A Survival Guide," Packt Publishing, 2017. https://learning.oreilly.com/library/view/penetration-testing-a/9781787287839/index.html.
6. Infosec Institute. (2016): "Pros and Cons in Penetration Testing Services: The Debate Continues," November 30, 2016. https://resources.infosecinstitute.com/pros-and-cons-in-penetration-testing-services-the-debate-continues/#gref.
7. Curry S. (2020): "We Hacked Apple for 3 Months: Here's What We Found," Sam Curry Blog, October 7, 2020. https://samcurry.net/hacking-apple/.

CHAPTER 18: CONCLUSION

1. Statistia. (2020): "Number of Smartphone Users Worldwide from 2016 to 2020," Statista. https://www.statista.com/statistics/330695/number-of-smartphone-users-worldwide/.
2. Statistia. (2020): "Number of Internet of Things (IoT) Connected Devices Worldwide in 2018, 2025 and 2030," Statista. https://www.statista.com/statistics/802690/worldwide-connected-devices-by-access-technology/.

Index

Printed in the United States
by Baker & Taylor Publisher Services